Mathematics for the Majority

# Algebra of a Sort



Chatto and Windus Educational
for
The Schools Council 1973

Granada Publishing Limited
First published in 1973 by Chatto and Windus Educational
Frogmore St Albans Hertfordshire AL2 2NF
and
3 Upper James Street London W1R 4BP



ISBN 0 7010 0617 X



Text set in 10/11 pt IBM Univers, printed by photolithography, and bound in Great Britain at The Pitman Press, Bath

# Contents

# Mathematics for the Majority

The Schools Council Project in Secondary School Mathematics (now called *Mathematics for the Majority*) was set up to help teachers construct courses for pupils of average and below-average ability, to relate mathematics to their experience, and to provide them with some insight into the processes that lie behind the use of mathematics as the language of science and as a source of interest in everyday things.

Members of the Project Team (1969–72):

P. J. Floyd (Director)
J. H. D. Parker (Deputy Director)
T. M. Murray-Rust
E. T. Norris
Mrs J. Stephens
P. A. Kaner (Evaluator) (to 1971)
M. J. Cannon (Evaluator) (from 1971)

The chief author of this book is:

E. T. Norris

# 1 What is algebra?

The memory which most teachers probably have of school algebra is of solving equations, factorizing expressions, removing brackets, simplifying products and quotients, and so on. It was all rather confusing and meaningless, and many of us looking back might find it difficult to give a satisfactory answer to the question: 'What is the use of algebra?' or more fundamentally: 'What is algebra?'

## Algebra as the study of generalizations

There are many different kinds of algebra. The symbols, formulae and equations of chemistry constitute an algebra. 'Modern' mathematics provides us with the algebra of sets and the algebra of electrical circuits. This book is concerned with traditional algebra, which can perhaps best be defined as the study of generalizations. In school mathematics such generalizations are often stated in symbols and in formulae, but in the early stages pupils should understand and express them in *words*. Only much later does it become appropriate to put them into symbolic form.

If this basis of algebra is accepted, we see at once that at many points in the pupil's mathematical education, from early years onwards, the teacher should be on the lookout for opportunities to help the pupil to think in general terms about what he is doing and to put these generalizations into words. In fact it can be argued that an important objective of education is to enable pupils to see generalizations in the mass of particular facts and experiences which they meet both in and out of school. Alfred North Whitehead asserted that 'the aim of scientific thought is to discover what is general in what is particular', and much of our thinking should be 'scientific' even when it is not applied to science. For instance in geography a pupil learns at first about the specific climatic conditions in one country and about the crops which grow in that country. Later he comes to understand that if the same conditions are found in another country, the same crops can probably be grown there. He has made a generalization. To take a more simple illustration, he sees the sun rise towards the east morning after morning and he comes to feel sure that this is what it will continue to do.

## Learning from experience

We often call the process of drawing conclusions from a set of observations 'Learning from experience'. Provided that this is understood to include many experiences which teachers select in order to lead the pupil to certain general conclusions, this should be a major aim of education and an aim for every teacher in his own particular subject.

We must adopt this aim in our mathematics teaching. Individual facts are important, but their relationship to other facts is far more important. For example, a young child discovers that three and four make seven. That is one fact, and to him it is a new and different fact that four and three make seven. Years may pass before he comes to understand that any pair of numbers can be added in either order to give the same result. Yet this is a far more important discovery, even though he may find it difficult for a long time to express it in words. It is the same with multiplication; primary school teachers know that children may quote and use certain 'bonds' long before it dawns on them that, say, six eights are necessarily the same as eight sixes.

## The reasoning we call 'inductive'

The examples we have just given illustrate 'inductive' generalizations, a descriptive term which covers all conclusions which we reach as a result of many observations. The conclusions of the geographer, economist, sociologist, and so on, and most of the 'laws' of the scientist, are *inductions* from particular cases. The truth of each assertion depends on the increasing weight of evidence provided by more and more instances or experiments which are 'favourable'. To take an illustration from prime numbers:

8 is the sum of two primes 3 and 5;
24 is the sum of two primes 7 and 17;
50 is the sum of two primes 13 and 37.

Many similar examples make one fairly convinced that it is true that 'every even number greater than 2 can be expressed as the sum of two primes'. Further testing strengthens this conviction.

## Patterns may mislead

If we find a single example which contradicts a tentative generalization, or *hypothesis,* this of course is fatal to it. If for instance we substitute any number for $n$, from $n = 1$ up to $n = 40$, in the expression $n^2 - n + 41$, we obtain a prime number.

$$n = 1 \Rightarrow n^2 - n + 41 = 41$$
$$n = 2 \Rightarrow n^2 - n + 41 = 43$$
$$n = 3 \Rightarrow n^2 - n + 41 = 47$$
$$\cdots\cdots\cdots\cdots\cdots\cdots\cdots$$
$$n = 40 \Rightarrow n^2 - n + 41 = 1601$$

(The arrow $\Rightarrow$ is the *implication* sign, which can be read as 'implies that'.)

By this time it seems highly probable that here we have a rule for obtaining prime numbers, but if we put $n = 41$, then we find that the 'rule' breaks down:

$$n = 41 \Rightarrow n^2 - n + 41 = 1681 = 41 \times 41.$$

Another striking illustration of the danger inherent in making a generalization from a mounting weight of evidence is taken from chapter 3 of *Mathematics in Primary Schools,* (ed. Association of Teachers of Mathematics, CUP).

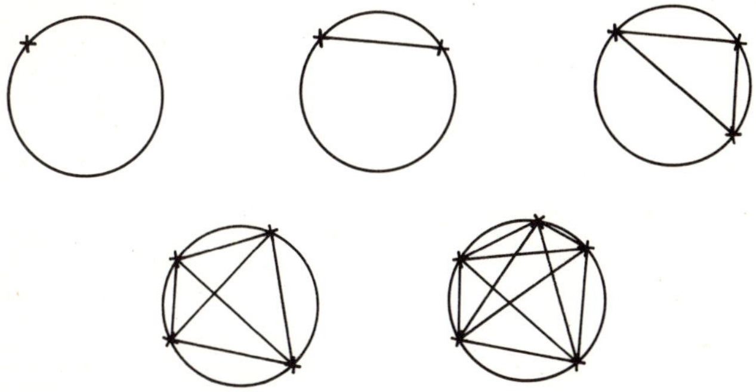

Fig. 1

We take in succession 1, 2, 3, 4, 5 points on a circle, regularly or irregularly spaced, and count the number of regions into which the circle is divided when we join pairs of points in all possible ways (fig. 1).

The numbers of regions are successively 1, 2, 4, 8, 16. It looks very much as if the circle with six points will give 32 regions but this is not so. The pattern suggests an inductive generalization which is false.

**Situations which can be analysed**

We see that the generalizations we make by induction from a number of cases may have a high *probability* of truth, but they need continually to be tested, and a single *counter-example* will be enough to prove them false. There are other generalizations, however, whose truth does not depend on the overwhelming evidence from a number of cases, although it may be that the study of a number of such cases will suggest to us the likely truth of a generalization in the first place. This happens when we analyse a situation and discover each time a common feature related to the *process,* and which is independent of the particular numbers involved.

For example, a pupil may have discovered how to measure the area of a rectangle by counting the number of (unit) squares which will fit inside it. If the rectangle measures 6 cm by 4 cm he will find that he can put a row of 6 cm squares in the rectangle along its longer side, then a second row of six, and eventually four such rows.

> The 'proof' of this is too difficult for the average pupil to attempt, although all pupils should recognize that its truth can be seen in the pattern of the diagrams in fig. 2. It will often be the case that a hypothesis is easy to formulate by inductive reasoning but is difficult to *prove.*

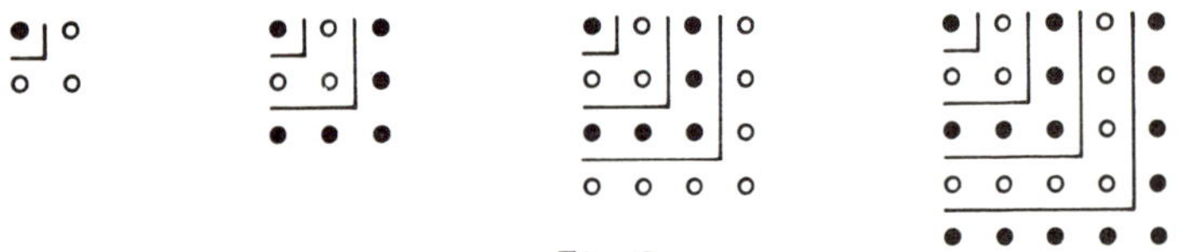

Fig. 2

A further example is that of the pattern:

$$\begin{aligned} 1 &= 1 = 1^3 \\ 3 + 5 &= 8 = 2^3 \\ 7 + 9 + 11 &= 27 = 3^3 \\ 13 + 15 + 17 + 19 &= 64 = 4^3 \end{aligned}$$

where successive groups of consecutive odd numbers, one extra being taken each time, give a sequence of cube numbers. In such cases the important activity is that of observing a pattern and formulating a rule.

We emphasize again that there are two kinds of generalization, the one whose truth we can only state with a high degree of probability, and the one whose truth we prove by analytical reasoning, provided we know how. Since each kind can usually only be discovered at first through observing a pattern in a large enough number of examples and because the proof of most hypotheses of the second kind may be beyond the abilities of our pupils, we shall ignore the distinction in the rest of this book, and shall take few generalizations further than the inductive reasoning stage. We assure the reader, however, that in the examples which we subsequently discuss in this book, and in which we appeal to inductive reasoning, proof by analysis is always possible.

A little more practical experience of doing this may, perhaps with teacher prodding, lead him to the conclusion that, no matter what numbers are involved, the area is always the result of multiplying the length number, or how many squares can be placed in a row, by the breadth number, or how many rows there will be. The pupil has made a generalization from a number of examples, but its truth does not become more certain as he considers more and more cases. The 'analysis' of a single example can lay bare the essential process and so establish the truth of the generalization.

**Inductive reasoning and analysis**

We can illustrate the distinction between our two kinds of reasoning, in two examples.

(a) The hypothesis that every even number greater than 2 is the sum of two primes depends on evidence accruing from more and more particular cases. No one has ever established its truth by analysis. It remains a generalization resting on the weight of evidence.

(b) We add 1 and 3; 1, 3 and 5; 1, 3, 5, and 7; 1, 3, 5, 7, and 9; and obtain successively 4, 9, 16, 25. Most of us, teachers as well as pupils, would now be prepared to guess that the 'next' set of odd numbers, up to 11, would give 36, the next square number, and that those up to 13 would give 49, and so on. That is to say, we are making the *hypothesis* (generalization) that 'the sum of any number of consecutive odd numbers, beginning with 1, will be a square number'. This hypothesis can be shown to be true by analysis, and so is an example of our second type of generalization.

**A note on inductive and deductive reasoning**

The inductive reasoning we have been talking about, and which we sometimes refer to as *induction,* is totally different from the method of reasoning known as *mathematical induction,* and the reader is asked to bear this fact in mind.

The process of inductive reasoning is diametrically opposed to that of deductive reasoning, which is reasoning by logical inference from one step to another. Both processes are mathematically respectable, and are essential tools in the search for knowledge. The teacher must be aware of the existence of both and should know when it is appropriate to use either one or the other. We believe that as a means of learning, inductive reasoning is encouraged far less than it ought to be, and that deductive reasoning is correspondingly over-stressed. Very often teachers feel that anything which falls short of rigorous deduction is not mathematically (or educationally) 'respectable', and consequently they tend to neglect any appeal to intuition in, for example, discovering the properties of shapes. This is a mistake. Many mathematical geniuses have made their discoveries through inductive reasoning, and as a tool it should be recognized as a most valuable initial means of discovery and learning in much of the arithmetic and algebra which we teach in school.

### Inductive reasoning in arithmetic

The inductive, or 'it looks as if', approach applies very often as much in arithmetic as it does in algebra. We have already quoted the apparently commutative nature of addition and multiplication when we are dealing with numbers, accepted as a 'law' simply because it holds for every case we have tested. 'It looks as if the product of a pair of numbers is the same whatever the order in which they are taken'. We have not 'proved', or deduced, that it will be true for every possible pair of numbers.

Another example is that of the equivalence of fractions. Pupils learn from a variety of practical activities, using rulers, diagrams, fraction boards and other aids, that certain pairs of fractions are equivalent to one another, e.g.

$$\frac{2}{3}=\frac{8}{12} \quad \frac{3}{4}=\frac{6}{8} \quad \frac{4}{10}=\frac{2}{5} \quad \frac{3}{4}=\frac{9}{12}$$

and so on.

With a little help perhaps from the teacher, the pupil comes to realize that multiplying or dividing the top and bottom parts of a fraction by the same number always gives an equivalent fraction. We see here how a succession of practical discoveries comes to form the basis for an inductive generalization, which then renders further practical exploration unnecessary.

Pupils can discover many of the processes of arithmetic by inductive reasoning, and this is far more effective than trying to remember teacher-given rules. One of the intentions in subsequent chapters of this book is to show how inductive reasoning can be encouraged in the learning of algebra. There is much lip service paid to the dictum 'proceed from the particular to the general' but algebra courses as a rule do not seem to subscribe to it.

### An example of inductive reasoning in algebra

If we arrange 16 pegs in a square array on a pegboard, we have a four-by-four square (fig. 3). If we remove the pegs in the bottom row of this square and use them to make a new column on the right hand side of the square, we find that we now have three rows of five with one peg left over.

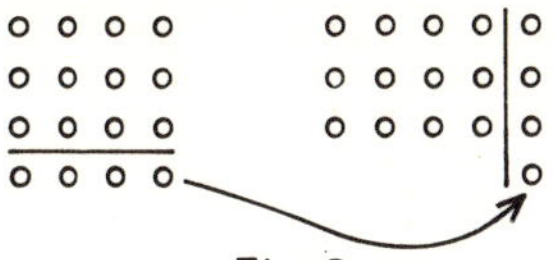

Fig. 3

We can do the same with a two-by-two square, a three-by-three square, a five-by-five square, which will give us one row of three plus one, two rows of four plus one, and so on. We then ask our pupils to guess how we have made the patterns and what the 'next' one is likely to be (fig. 4).

Fig. 4

The *numerical* pattern here is

$$1 \times 3 + 1$$
$$2 \times 4 + 1$$
$$3 \times 5 + 1$$
$$4 \times 6 + 1$$

We can expect our pupils to say: 'It looks as if the next one will be $5 \times 7 + 1$ and the one after that $6 \times 8 + 1$'. In doing this they are using inductive reasoning.

When the pupils count the pegs in each arrangement they find that there are successively 4, 9, 16 and 25. They possibly recognize these as square numbers and guess that the 'next' totals ought, therefore, to be 36 and 49 respectively. A check show that:

$$5 \times 7 + 1 = 36$$
$$6 \times 8 + 1 = 49$$

Since $9 = 3^2$ and $16 = 4^2$, our pupils should finally be able to say that the patterns on the board were made from squares, and they can find out for themselves how moving the pegs from the bottom row of each square to a new position at the right hand side of the other columns gives an extra column with a peg left over.

If we look again at the numerical pattern, thinking of 9 as $3^2$, 16 as $4^2$, and so on, we have:

$$2 \times 4 + 1 = 3^2$$
$$3 \times 5 + 1 = 4^2$$
$$4 \times 6 + 1 = 5^2$$
$$5 \times 7 + 1 = 6^2$$
$$6 \times 8 + 1 = 7^2$$

We can then make some *general* statement as:
'The product of two numbers, one of which is 2 more than the other, with one added, gives the square of the number between them'.

We can write this statement in a mathematical shorthand provided we know how to translate phrases like 'a number which is two more than another', and so on, into symbols. It may help to refer to a number line (fig. 5).

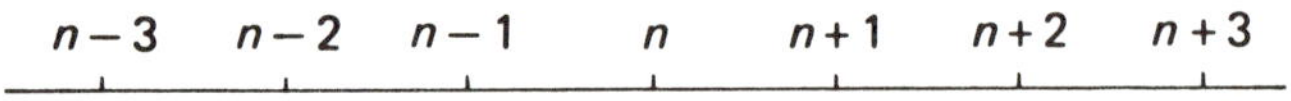


Fig. 5

If we take our two numbers as $n$ and $n + 2$, the number between them will be $n + 1$, and so we can write:

$$n \times (n + 2) + 1 = (n + 1)^2$$

or, multiplying according to the rules of algebra (see page 59):

$$n^2 + 2n + 1 = (n + 1)^2$$

which is a familiar algebraic identity usually derived in quite a different way, and almost certainly not from a number situation like this. (Is this true for the consecutive numbers $(n - 3)$, $(n - 2)$ and $(n - 1)$? Or for *any* three consecutive numbers?)

If we reverse our verbal statement, and say: 'The square of a number is equal to one more than the product of the two numbers immediately on either side of it', this translates into symbols as:

$$n^2 = 1 + (n - 1) \times (n + 1)$$
$$\text{or } n^2 = 1 + (n - 1)(n + 1)$$
$$\text{or } n^2 - 1 = (n - 1)(n + 1)$$

which is familiar to teachers, and is a special case of the 'difference between two squares'.

Once we have formed a generalized statement using symbols, we should next find out whether the generalization is true for all numbers, and not only for those whose behaviour has allowed us to make such a tentative statement. The statement:

$$n^2 + 2n + 1 = (n + 1)^2$$

is true for the set of natural numbers; is it also true for the set of ratio numbers? (See *Number Appreciation,* chapter 3.)

If we put $n$ equal to $\frac{1}{2}$, we find that the statement holds:

$$\text{L.H.S.} = (\tfrac{1}{2})^2 + (2 \times \tfrac{1}{2}) + 1 = \tfrac{1}{4} + 1 + 1 = 2\tfrac{1}{4}$$

$$\text{R.H.S.} = (\tfrac{1}{2} + 1)^2 = (1\tfrac{1}{2})^2 = 2\tfrac{1}{4}$$

and the left hand side is equal to the right hand side. Further trial with other fractions strengthens our conviction that the statement holds in the set of rationals.

This illustrates an important property of a generalization, that it often widens the application of a particular rule in ways which otherwise might have remained unsuspected.

## Some advantages of the inductive approach

The foregoing section gives one example of how we can use inductive reasoning (for there has been no *deduction*), first to make general statements in *words* about the relationships we see in number patterns, and then to translate these into statements using *symbols,* which are the essence of algebra. The advantages of such an approach are:

(a) Number patterns are usually interesting in themselves.
(b) The generalizations they are able to make in this way will have more meaning for our pupils than if presented in an abstract way, which is often the textbook approach.
(c) Practice in making generalizations and in translating them from 'English language' into 'Mathematics language' and vice versa, leads to a deeper understanding of the meaning and use of symbolic statements or *formulae.*
(d) The ability to make generalizations contributes to the mental maturing of our pupils.

In making generalizations, whether they are simple ones such as that 'addition of numbers is commutative ($a + b = b + a$)', and that 'if one acute angle of a right-angled triangle is $A^\circ$, the other is $(90 - A)^\circ$', or more complex ones such as those we have just been considering, we move informally and incidentally into the realm of algebra. It is important that we should do this, because in becoming familiar with the language and operations of algebra our pupils will develop certain powers of expression and thought, which are the real objectives we are striving for in all this work; while for some of our pupils this kind of experience can form the basis of a later more systematic study of algebra.

## The use of formulae

Mathematical generalizations can often be expressed as algebraic formulae. Such formulae, more than any other mathematical tool, allow us to apply mathematical ideas to real situations and so make the link between theoretical principles and practical needs and interests, which may go some way towards convincing our pupils of the value of algebra.

One of the great names in mathematical education is T. P. Nunn, who wrote many years ago that 'mathematical truths always have two sides. With one they face and have contact with the world of realities in time and space; with the other they face and have relations with one another'. (A simple illustration is the use of the 3, 4, 5 triangle to lay out a right-angle on a building site. Here is the contact of a mathematical truth with the world of reality, while at the same time the appearance of this truth reminds us of all the theoretical implications of the Pythagorean theorem.) Thus practice and theory in mathematics are two facets of intellectual activity which should not be divorced from each other. They have run side by side throughout the development of civilization, and mathematical education involves the appreciation of this dual role. Nunn stresses that in a school mathematics course the two aspects, practical and theoretical, can and should be developed together, and that by choosing material which shows clearly the practical value of mathematics, be it algebra, geometry or any other branch, we shall find the best opportunities for developing the theory of the subject. In algebra the formulae used by the engineer or by the technician show the practical value of algebraic symbolism, while the use and manipulation of formulae bring us face to face with the relationships in which the theoretical structure of mathematics is revealed.

### Other components of algebra

When, however, we come to look at the traditional items of an algebra syllabus, at the chapter headings of many textbooks, or at the questions found in some examination papers, we often find little indication that formulae sound the dominant note. What evidence there is seems to suggest that the major parts of the syllabus are:

1. Conventions and signs used in symbolic notation.
2. Meaning and use of indices.
3. Substitution.
4. Collection of terms and simplification of expressions.
5. Use and removal of brackets.
6. Factors.
7. Solution of equations.

Even when the syllabus puts early emphasis on formula-making, this is rarely followed up, except perhaps in 'changing the subject of a formula' where problems are usually isolated and out of context.

In *From Counting to Calculating* the point is made that one of the reasons for teaching arithmetic is for its applications in life. Similarly one of the reasons for teaching algebraic techniques is to enable our pupils to understand algebraic formulae and to use them effectively. We now look briefly at examples from the major parts of the syllabus which we listed above, to gain some idea of what these techniques are.

1. *Conventions*

A little thought about vehicles moving at constant speeds leads us to conclude that 'the distance travelled is the speed multiplied by the time'.

Using initial letters to represent *quantities*, we can write this statement concisely as:

$$d = s \times t$$

After a little practice at doing this with other quantities and other statements, we can economize by leaving out the multiplication sign:

$$d = st$$

with the understanding that letters written side by side are to be multiplied. The same rule will apply to a mixture of letters and numbers, with the added convention that when the multiplication sign is omitted the number is written first.

e.g. 'the diameter of a circle can be found by multiplying the radius by 2',

leads to $d = r \times 2$
or $d = 2r$ (not $d = r2$).

2. *Indices*

What we know about the area of a square and the volume of a cube can be expressed as $A = l \times l$ and $V = l \times l \times l$. By our convention above these can be written as:

$$A = ll \text{ and } V = lll$$

which were the methods used in the seventeenth century, when a satisfactory algebraic symbolism was being evolved. (It is useful for the teacher to know something of the historical development of mathematics; it often gives important clues to difficulties met by our pupils and to ways in which topics might be introduced and developed.) The next step is to adopt the further shorthand of the standard index notation, $A = l^2$, $V = l^3$, although the first acquaintance with this notation is probably better made through number situations, such as $2 \times 2 \times 2 = 2^3$, to reduce the possible confusion between say, $l^3$ and $3l$.

3. *Substitution*

Calculations with actual formulae taken from collections used by engineers, technicians, etc., have point and meaning. For example 'The greatest safe load ($L$) for a certain type of chain is given by the formula $L = 1{\cdot}4d^2$ ($L$ in metric tonnes, $d$ the diameter in centimetres); find the safe load for a 4 cm chain.' This is mathematically the same problem as 'find the value of $xy^2$ when $x = 1{\cdot}4$, $y = 4$'; but whereas the first problem makes some kind of sense, being a real one for somebody, if not for our pupils, the second is clearly good for nothing except giving practice in substitution.

4. *Collection of terms*

A very simple concrete example:

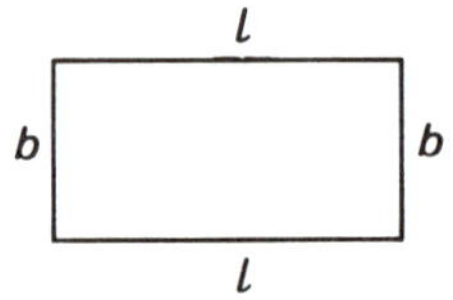


Fig. 6

The formula for the perimeter of a rectangle may be expressed as:

Perimeter = length + breadth + length + breadth

or

Perimeter = length + length + breadth + breadth

or

Perimeter = twice the length + twice the breadth.

These statements in symbols become

$$p = l + b + l + b$$

or

$$p = l + l + b + b$$

or

$$p = 2l + 2b.$$

It is essential that we approach the collection of terms in some such way as this, otherwise our pupils will continue to find it difficult to avoid the pitfall of adding $3a + 2a$ as $5a^2$. Also it may be as well to write $b$ as $1b$ to begin with: pupils have been known to treat $b + 2b$ as being incapable of further simplification. This is quite reasonable: we know that *numbers* can be added, whereas we may not be quite so certain about when we may add *letters.*

5. *Brackets*

We can again use the perimeter of a rectangle, stating our formula initially in words if we think this necessary.

Halfway round is $l + b$ (fig. 6), so the whole perimeter (twice this) is $2 \times (l + b)$, or using the convention which allows us to drop the multiplication sign:

$$p = 2(l + b).$$

Many practical formulae, of course, contain brackets and gain in conciseness as a result.

6. *Factorizing*

This can be justified whenever it reduces the number of calculations which we need to make when we substitute in a formula. Factorizing involves the use of brackets (see above).

For instance the area ($A$) of an annulus (ring) is given by

$$A = \pi a^2 - \pi b^2$$

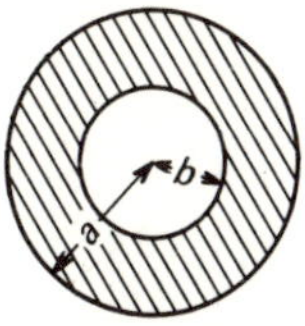


Fig. 7

where $a$ and $b$ are the outer and inner radii (fig. 7). To find the area from this formula we have to perform five computations – four multiplications and one subtraction.

Using the common factor, the formula becomes

$$A = \pi(a^2 - b^2)$$

which reduces the number of computations to three multiplications and one subtraction.

In the form

$$A = \pi(a + b)(a - b)$$

the number of computations is now two multiplications, an addition and a subtraction.

We see here, as we did in 4 and 5, how factorizing and the use of brackets can reduce the number of operations needed for finding the numerical value of a formula in which we have substituted numbers. This often simplifies the calculation considerably and we should help our pupils to judge when such simplification is likely to lead to a saving in time and effort.

7. *Solution of equations*

It is a relatively simple matter to use a formula in a 'forwards' direction. For example, we can imagine an object shot vertically into the air at a velocity of 200 m/s. If we are asked to calculate its height after 3 s, and we are given the formula $h = 200t - 4{\cdot}9t^2$, all we have to do is to substitute $t = 3$ in the formula.

$$\begin{aligned} h &= 200t - 4{\cdot}9t^2 \\ &= (200 \times 3) - (4{\cdot}9 \times 3 \times 3) \\ &= 600 - 44{\cdot}1 \\ &= 555{\cdot}9 \end{aligned}$$

Its height is about 556 m.

If, on the other hand, we are asked, not 'given the time, find the height' but 'given the height, find the time', the situation is quite different. When, for example, would the object be 500 m above the ground?

In this case we still substitute, but this time we find we have to solve a quadratic equation:

$$500 = 200t - 4{\cdot}9t^2$$

and it may be that the study of such a situation can point the general need for devising some method of solving quadratic equations.

As a further example, we may be asked: 'What is the Fahrenheit temperature when the Centigrade reading is 20°?' All we do is to substitute $C = 20$ in the formula.

$$F = \frac{9}{5}C + 32$$

$$\begin{aligned} F &= \frac{9}{5} \times 20 + 32 \\ &= (9 \times 4) + 32 \\ &= 68. \end{aligned}$$

All formulae give direct numerical answers when used forwards, but when used backwards they give rise to equations. For the question: 'What is the Centigrade temperature when the Fahrenheit reading is 68°?' our formula

$$(F = \frac{9}{5}C + 32)$$

will give the equation:

$$68 = \frac{9}{5}C + 32$$

or $$\frac{9}{5}C + 32 = 68.$$

In principle this is no different from, say,

$$4x + 19 = 75$$

which is no more than a contrived example designed to give practice in solving equations.

The main problem comes when the pupil has actually to find the correct number to replace $C$ in the equation

$$\frac{9}{5}C + 32 = 68,$$

i.e. to solve the equation.

The above examples, some fairly trivial, some not, show how the traditional topics and techniques of algebra are directly related to formulae, and why they must be mastered by anyone making a systematic study of the subject.

For our pupils the fundamental aim is to develop the power of generalization. The extent to which this power is channelled into symbolism and its subsequent transformations (many of which, indeed, prove to be not merely mechanically advantageous but even creative of more mathematics) is a matter for choice and decision by teachers. It is recognized here that some of the pupils under consideration will enter for public examinations, which for the time being require parts of traditional formal algebra; consequently more detailed suggestions are included in subsequent chapters which go beyond the range of work deemed generally suitable. The material may also serve to indicate ways of development which could be used for brighter pupils as well. Quite rightly, some of the algebra which the majority of the teachers of the present day themselves learned is being set aside in favour of more relevant topics; this can be welcomed, as long as these are not treated in as formalistic a way as the old topics often were, and above all, so long as generalization is recognized to be the underlying basis. Nevertheless, in various fields of science and in other branches of mathematics, the notation, structures and methods of traditional algebra continue to be needed; when they are made relevant and understandable they still have a part to play in the full education of most of our pupils.

# 2 Induction and generalization from patterns

In this chapter we shall give illustrations to show how we may detect patterns in relationships involving numbers, from which we can make inductive generalizations.

We may identify three stages in such a programme.

1. In most of the examples the patterns are interesting in themselves and could be used with pupils on a purely numerical level without attempting to make a conscious generalization. Pupils would predict the 'next' numbers in a pattern. In doing so they would show that they had perceived the pattern and made an induction from it.
2. Pupils would be encouraged to express in words the relationship they had discovered in the pattern, and then to test the truth of their assertions for new particular cases. For many pupils this might be as far as they can go.
3. Some pupils would be able to write their verbal generalizations in a shorthand form, using symbols. The symbolic rule would be checked by a substitution to confirm cases already noted, and to predict later numbers of the pattern which would also be checked. The rules of formulae thus expressed will rarely have any value in themselves for our pupils, and there should be no questions of their being memorized; *their importance lies in the processes of thought by which they have been reached.*

It will happen rarely, and only with very mature pupils, that the rule reached by induction (the result of 'it looks as if . . .' thinking) will be established by deduction and logical inference. Teachers will appreciate that this is the crown and fully satisfying end of mathematical thought; nevertheless, there is satisfaction and often excitement to be derived from each of the earlier stages which for our pupils is undoubtedly 'real' mathematics.

A few further points need to be emphasized in connection with the examples which follow. Firstly, they are of varying degrees of difficulty, so that those which the teacher deems at all suitable for pupil investigation and discussion will be used with pupils of widely differing ages and abilities. Secondly, they are not presented in a sequential development of content, and are not offered as a preparation for the work which we shall discuss later. Many of the ideas and techniques developed, for example, in Chapter 3 on formulae, must be part of a pupil's mathematical background before he can devise and use the symbolic notation which expresses generalizations previously stated in words. Thirdly, the three stages referred to above should not follow one another in rapid succession until pupils have had considerable experience of manipulating algebraic expressions. There should normally be a considerable time allowed for mental assimilation between one stage and the next (such as the wise teacher gives in the development of any topic) especially before symbolism is introduced. In any case 'stage three' in some of the examples may not be suitable for all of our pupils, but it is included here in order to round off for teachers the discussion of a particular pattern. Finally, some of the illustrations appear in other books, such as *Mathematical Pattern,* where different aspects of their importance are considered.

**Example 1**

In *Machines, Mechanisms and Mathematics* it is suggested that pupils investigate the properties of polygonal linkages made from a number of struts. Pupils should discover that in order to make these linkages rigid we can insert *diagonal* struts which divide the polygons into a number of triangles.

By collating the information for a number of polygons, pupils may compile the following table:

| | | | | |
|---|---|---|---|---|
| Number of sides of polygon | 3 | 4 | 5 | 6 |
| Number of triangles formed | 1 | 2 | 3 | 4 |

Without needing to make polygons of seven and eight sides we may say that 'it looks as if the number of triangles formed will be five and six respectively', and these expectations can then be checked. This is the first stage of inductive reasoning.

The second stage directs attention to the relationship perceived and will yield the verbal generalization that 'the number of triangles formed is two less than the number of sides'. The third stage then asserts that in a polygon of $n$ sides the number of triangles formed will be $n - 2$.

When the enquiry is directed further to the angle sum of irregular polygons, starting from the fact that the sum of the angles of a triangle is two right-angles, examination of particular polygons will yield the following data:

| | | | | |
|---|---|---|---|---|
| Number of sides of polygon | 3 | 4 | 5 | 6 |
| Number of triangles formed | 1 | 2 | 3 | 4 |
| Sum of angles (in right-angles) | 2 | 4 | 6 | 8 |

Inductions for the angle sum of seven-, eight-, . . . sided polygons follow from the pattern perceived.

In words, we can say that 'the number of right-angles is twice the number of triangles', or – going a step further – 'take two from the number of sides and then double'.

It must be stressed that for pupils – especially for those with whom we are mainly concerned – to reach the stage of being able to generalize in words in this way is a significant achievement. It is a step they should be encouraged and helped to make, and for some it may be as far as they can and should go. But they have seen a pattern and then a relationship, and have expressed what they see in a generalization; and this is genuine mathematical thinking.

For those who can go further it is not wise to hurry into symbolism, although for the teacher it may seem so simple. It is best to allow the verbal generalization to be appreciated, used, thoroughly understood and accepted before going further. After an interval (it might be measured in weeks for some pupils, in months for others) the symbolic notation can be introduced as a shorthand. The 'translation' of the sentence would be that the number of right-angles in all the angles of a polygon of $n$ sides is $(n - 2) \times 2$ or $2(n - 2)$. This too will be appreciated as a meaningful condensing of what has previously been expressed in a much more long-winded way; but it should be emphasized that the symbolic rule itself is of little importance and is probably better not used in that form. It may be pointed out that this formula is a direct translation of the verbal generalization, and is more significant than the conventional '$2n - 4$' right-angles.

**Example 2**

Is it true that the angle in a semi-circle is a right-angle?

Take a succession of numerical values for angle $q$ (fig. 8), for example, let $q$ be $20^\circ$.

Then by the isosceles triangle property, $x = 20^\circ$;

by the external angle property, $c = 40^\circ$;

by the angle-sum property, $r + y = 140^\circ$;

but $r = y$ in the isosceles triangle CPR, so $y = 70^\circ$.

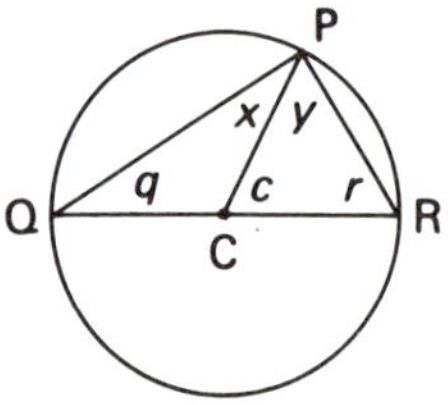


Fig. 8

Hence $P = x + y = 20^\circ + 70^\circ = 90^\circ$

Let pupils make further similar calculations, each choosing a different value for $q$; then build up a table, as below:

| | | | | |
|---|---|---|---|---|
| $q$ | 20° | 28° | 35° | |
| $x$ | 20° | 28° | 35° | |
| $c$ | 40° | 56° | 70° | |
| $r + y$ | 140° | 124° | 110° | etc. |
| $y$ | 70° | 62° | 55° | |
| $P = x + y$ | 90° | 90° | 90° | |

In all cases examined the angle $P$ is a right-angle, so 'it looks as if' this is a property of a semi-circle.

Teachers will perceive that this example of generalization is clearly based on *analysis* rather than on the overwhelming evidence of a large number of cases – the distinction referred to in chapter 1. It may not appear in this light to our pupils who will be more impressed perhaps by the constancy of the results. Only much later may they come to see, on looking at the property again, that an analysis of the situation would suggest that we could go on to establish the general relation by deduction.

**Note:** This exercise is useful in developing the idea that the symbol $q$ represents a *variable* (within a certain range of values). The symbols $x, y, c, r$ are also found to represent *variables* and yet surprisingly $x + y$ has a *constant* value.

**Example 3**

(1) In chapter 1 the following sequence was investigated:

$2 \times 4 + 1$
$3 \times 5 + 1$
$4 \times 6 + 1$ etc.

and various inductions were made. (See pp. 7, 8 and figs 3, 4.)

(2) Analogous patterns can be seen in the following numbers:

$2 \times 6 + 4$
$3 \times 7 + 4$
$4 \times 8 + 4$
$5 \times 9 + 4$ etc.

which describe the arrangement of pegs on a pegboard after *two* rows of a square have been changed into columns (fig. 9).

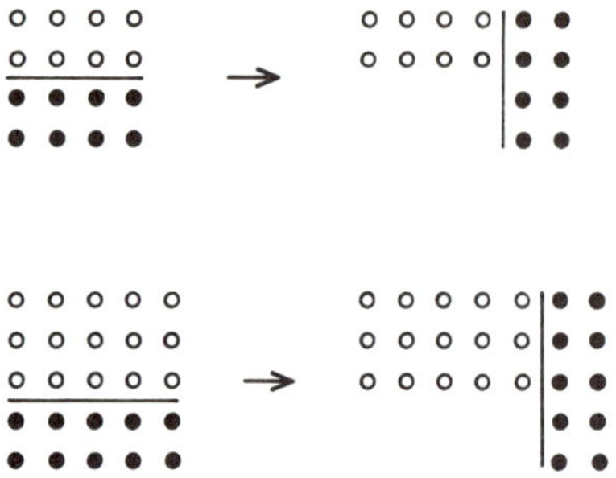

Fig. 9

From the pattern of numbers we can induce that:

(i) the 'next' line will be $6 \times 10 + 4$;

(ii) the result of each operation is to give a square number; and the 'next' will be $8^2$;

(iii) 'it looks as if' the product of two numbers separated by 4, with 4 added, gives the square of the number mid-way between the two numbers;

(iv) in symbols the generalization is

$$n(n+4)+4=(n+2)^2.$$

or

$$n^2+4n+4=(n+2)^2$$

Writing $a$ for $(n+2)$ and hence $(a-2)$ for $n$, we can express this as

$$a^2=(a-2)(a+2)+4$$

or

$$a^2-4=(a-2)(a+2)$$

each of these expressions being an algebraic identity familiar to teachers.

**Note:** an algebraic identity is a sentence which is universally true. In our example $a^2-4=(a-2)(a+2)$ we have a sentence which is true for all values of $a$, i.e. we have a universal truth or an *identity.*

On the other hand, were we to write

$$a^2-4=0$$

we recognize that this is not true for all values of $a$ and our previous algebraic knowledge will lead us to see that it is true in two cases only and untrue in every other case. It is true when $a = {}^{+}2$ and when $a = {}^{-}2$. This type of sentence of limited truth is called an *equation.*

An identity is then a universal truth and an equation is a limited truth. These are important but very different types of algebraic sentences. If at any time we need to make clear that such a sentence is an identity rather than an equation we may use a special symbol, $\equiv$. For example,

$$a^2 - 4 = 0 \text{ (Equation)}$$

$$a^2 - 4 \equiv (a-2)(a+2) \text{ (Identity)}$$

(3) A further pattern of the same kind as in (2) is seen in the following sequence, from which similar inductions can be made:

$$\begin{aligned} &2 \times \ \ 8 + 9 \\ &3 \times \ \ 9 + 9 \\ &4 \times 10 + 9 \text{ etc.} \end{aligned}$$

A more general pattern now begins to emerge from (1), (2) and (3) when they are generalized as below, and further examples can be invented.

| | | |
|---|---|---|
| (1) | $n(n+2) + 1 = (n+1)^2$ | $(a-1)(a+1) + 1 = a^2$ |
| (2) | $n(n+4) + 4 = (n+2)^2$ | $(a-2)(a+2) + 4 = a^2$ |
| (3) | $n(n+6) + 9 = (n+3)^2$ | $(a-3)(a+3) + 9 = a^2$. |

Teachers will then see (but it is not suggested that this would be appropriate for pupils) that the inductive generalizations which could be made might be expressed as:

| | | | |
|---|---|---|---|
| | $n(n+2p) + p^2 = (n+p)^2$ | | $(a-b)(a+b) + b^2 = a^2$ |
| i.e. | $n^2 + 2np + p^2 = (n+p)^2$ | i.e. | $(a-b)(a+b) = a^2 - b^2$. |

These 'derivations' are not, of course, put forward as the best means of obtaining the identities but as further illustrations of the way patterns can be investigated inductively and generalizations made.

**Example 4**

A variation of the above is to start with differences of square numbers and to look for patterns in the results.

(1) Simple calculation gives

$$\begin{aligned} 1^2 - 0^2 &= 1 \\ 2^2 - 1^2 &= 3 \\ 3^2 - 2^2 &= 5 \\ 4^2 - 3^2 &= 7 \end{aligned}$$

from which several inductions can be made.

One of these relates to the odd numbers on the right to the sum of the numbers on the left which are squared.

We see, for example, that in the third line 5 is the sum of 3 and 2. In the fourth line 7 is the sum of 4 and 3.

So we can write

$$1^2 - 0^2 = 1 + 0$$
$$2^2 - 1^2 = 2 + 1$$
$$3^2 - 2^2 = 3 + 2$$
$$4^2 - 3^2 = 4 + 3$$

and so on (fig. 10).

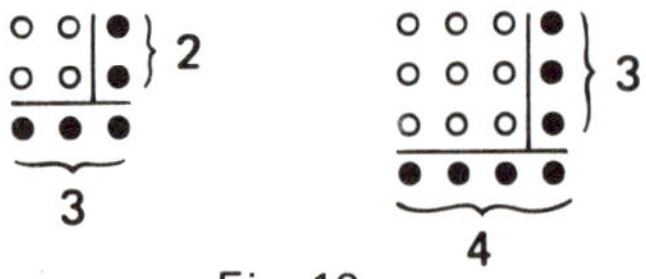


Fig. 10

(Then 'it looks as if' $12^2 - 11^2$ should equal . . .? and $101^2 - 100^2$ should equal . . .?)

(2) Suppose next that the difference in the numbers to be squared is two. Then

$$2^2 - 0^2 = 4$$
$$3^2 - 1^2 = 8$$
$$4^2 - 2^2 = 12$$
$$5^2 - 3^2 = 16$$

'and so on', with the numbers on the right increasing by four each time.

Can these equations be related to the sum of the numbers which are squared on the left, as in (1)?

Yes, because, reversing the order of the above equations gives:

$$5 + 3 = 8 \text{ and } 16 = 2 \times 8 = 2(5 + 3)$$
$$4 + 2 = 6 \text{ and } 12 = 2 \times 6 = 2(4 + 2)$$
$$3 + 1 = 4 \text{ and } 8 = 2 \times 4 = 2(3 + 1)$$

A pegboard interpretation of the pattern might be:

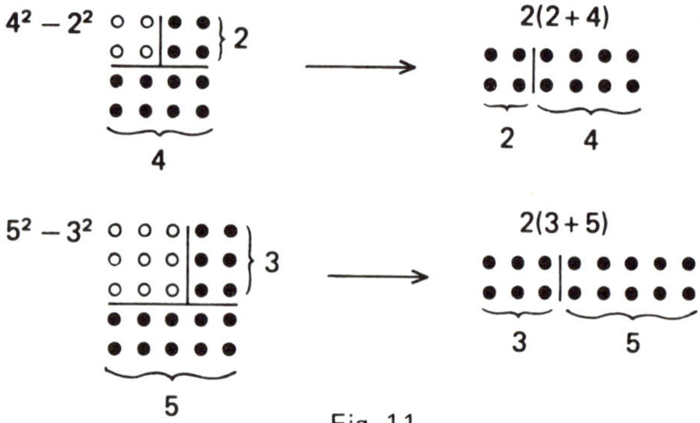


Fig. 11

So 'it looks as if' the next line $6^2 - 4^2$ should equal $2(6 + 4)$, and we find that this is so.

(3) If the left-hand numbers differ by three, then we find:

$$3^2 - 0^2 = 9$$
$$4^2 - 1^2 = 15$$
$$5^2 - 2^2 = 21$$
$$6^2 - 3^2 = 27.$$

But

$$9 = 3 \times 3 = 3(3 + 0)$$
$$15 = 3 \times 5 = 3(4 + 1)$$
$$21 = 3 \times 7 = 3(5 + 2)$$
$$27 = 3 \times 9 = 3(6 + 3)$$

Is the 'next' equation true?
That is, does $7^2 - 4^2 = 3(7 + 4)$?
(So what should $12^2 - 9^2$; $20^2 - 17^2$; $102^2 - 99^2$ equal?)

There is in the above outline far too strong a suggestion of a teacher-imposed direction to the investigations. Many teachers, however, have found that given only brief leads and hints the pupils themselves have made the discoveries, have suggested the extended developments, and have vied with one another in taking the patterns further and further and in testing whether their anticipations are correct.

Nevertheless, the teacher should help at least some pupils to see where the discoveries take them — to enable them to put their inductive conclusions into words, in fact, in such a statement as:

'In all these examples the difference of the squares of two numbers is the same as the difference of those numbers multiplied by their sum.'

If and when the time comes to put this conclusion into symbols, the translation is:

$$a^2 - b^2 = (a - b) \times (a + b)$$

There is a world of difference in mathematical understanding between a relationship discovered in such a way, and the mechanical multiplication of the brackets according to prescribed rules in order to obtain the left-hand side, justifiable though this procedure may be at a certain stage.

**Example 5**

Look at the following:

$$\left.\begin{array}{r} 1 . 2 . 3 . 4 + 1 = 25 \\ 2 . 3 . 4 . 5 + 1 = 121 \\ 3 . 4 . 5 . 6 + 1 = 361 \end{array}\right\} \text{where the dots denote multiplication.}$$

Pupils who have had plenty of experience with numbers quickly recognize 25 and 121 as perfect squares and will investigate 361 to find that it is the square of 19.

The pattern clearly goes on to

$$4 \,.\, 5 \,.\, 6 \,.\, 7 + 1$$

and 'it looks as if' this should give another square number. Can we forecast what it would be?

So far the numbers on the right-hand side of each line are $5^2$, $11^2$, $19^2$, and there is no obvious clue to the pattern.

Some pupils may see that the differences between 5 and 11 and between 11 and 19 are 6 and 8 respectively, and will guess that the next difference might be 10; the next square would then be $29^2$.

| | 5 | | 11 | | 19 | | (?29) |
|---|---|---|---|---|---|---|---|
| (differences) | | 6 | | 8 | | (?10) | |

And $29^2$ (= 841) proves to be correct.

(Note that if the pattern goes one step backwards it gives

$$0 \,.\, 1 \,.\, 2 \,.\, 3 + 1 = 1^2$$

which fits in with the pattern of differences of the numbers which are squared, for the previous difference ought to be, and is, 4.)

It does not help, however, in giving a general description of the pattern. We cannot immediately say, for example, what the 20th line would be.

Perhaps a few pupils will notice another relation – that the product of the first and last factors, with 1 added, gives the number to be squared.

$$1 \,.\, 2 \,.\, 3 \,.\, 4 + 1 = 25$$
$$1 \,.\, 4 + 1 = 5$$

$$2 \,.\, 3 \,.\, 4 \,.\, 5 + 1 = 121$$
$$2 \,.\, 5 + 1 = 11$$

and for the fourth line:

$$4 \,.\, 5 \,.\, 6 \,.\, 7 + 1 = 841$$
$$4 \,.\, 7 + 1 = 29$$

It is always stimulating, as well as being additionally convincing, when one line of inductive reasoning confirms another, and our pupils will often be intrigued enough to want to go on to test further items of the pattern.

Teachers may find other relationships between the numbers on the right-hand sides and square numbers, e.g.

$$5 = 2^2 + 1$$
$$11 = 3^2 + 2$$
$$19 = 4^2 + 3$$

so that 'it looks as if' the next number in the sequence will be: $5^2 + 4$.

This particular induction, moreover, enables us to make further progress. For if a 'stage two' generalization is desired, the statement would be: 'The product of four consecutive whole numbers, with 1 added, gives the square of one more than the product of the first and last of these numbers'.

(Can the range of numbers be extended? That is, does the assertion hold for fractional and directed numbers?)

Translation of the verbal generalization into symbols ('stage three') gives a striking demonstration that algebraic representation and manipulation establish firmly and concisely the truth of assertions such as the above which otherwise appear only as highly probable.

For the symbolization of the left is

$$n(n + 1)(n + 2)(n + 3) + 1$$

and of the right is

$$[n(n + 3) + 1]^2 \quad \text{or} \quad (n^2 + 3n + 1)^2.$$

On multiplying out these two expressions we see that they are identical, each being equal to $n^4 + 6n^3 + 11n^2 + 6n + 1$.

It need not be emphasized that very few of our pupils (if any) will be expected to reach this point in technique; but the demonstration illustrates to teachers the uses of manipulation.

The above discussion relating to the product of four consecutive integers may be linked with two other simpler 'product' patterns which offer opportunities for investigation.

(1) Products of two consecutive integers

$$1 . 2 = 1^2 + 1$$
$$2 . 3 = 2^2 + 2$$
$$3 . 4 = 3^2 + 3$$
$$4 . 5 = 4^2 + 4 \text{ etc.}$$

and (2) Products of three consecutive integers with an added term.

$$1 . 2 . 3 + 2 = 2^3$$
$$2 . 3 . 4 + 3 = 3^3$$
$$3 . 4 . 5 + 4 = 4^3$$
$$4 . 5 . 6 + 5 = 5^3 \text{ etc.}$$

**Example 6**

Notice the following relationships:

$$
\begin{aligned}
1 &= 1 \quad (= 1^3)\\
3 + 5 &= 8 \quad (= 2^3)\\
7 + 9 + 11 &= 27 \quad (= 3^3)\\
13 + 15 + 17 + 19 &= 64 \quad (= 4^3)
\end{aligned}
$$

On the left-hand side of each equation are groups of consecutive odd numbers, one more in each successive group; and on the other side are cube numbers.

'Stage one' then suggests the next line as

L.H.S. $21 + 23 + 25 + 27 + 29$
R.H.S. $125$ or $5^3$

and the values of these agree with each other.

An inductive generalization (stage two) is not easily made until we note, for example, that the number which is cubed on the right is the same as the number of terms on the left. We may also notice that the middle number in the third line is 9, or $3^2$, while the middle number in the fifth line is 25, or $5^2$. If we look at the fourth line we might expect to find 16; in fact there is no 'middle' number in the fourth line, but the *two* numbers in the 'middle' of this line are 15 and 17, which are, respectively, one less and one more than 16. If we notice that 13 and 19 are, respectively, three less and three more than 16, we may come to realize that 16 is the *average* of the terms on the left-hand side; this is $4^2$, and the number on the right-hand side is $4^3$. Similarly 25 is the average of the numbers in the fifth line.

Suppose now we wish to write the seventh line of the pattern. There will be seven consecutive odd numbers, and the middle one will be $7^2$ or 49, with three terms on either side.

We can write the relation down:

$$43 + 45 + 47 + 49 + 51 + 53 + 55 = 7^3 = 343.$$

In the tenth line, the average term must be $10^2$ or 100, and five consecutive odd numbers must be written on either side of it. Thus we find:

$$\begin{array}{c}(100)\\ \downarrow\end{array}$$

$$91 + 93 + 95 + 97 + 99 + 101 + 103 + 105 + 107 + 109 = 10^3 = 1000.$$

We note that 100 itself, being an even number, does not appear on the left hand side. Whenever the number of terms in a line is even their average will also be even. To write down the line, we square the average, and write down an equal number of consecutive odd numbers on either side of it. For example the 50th line will contain 50 numbers: these will be the 25 consecutive odd numbers immediately below $50^2$ or 2500, and the 25 immediately above it.

$$\ldots 2497 + 2499 + 2500 + 2501 + 2503 \ldots$$

The symbolic generalization (stage three) becomes fairly complicated, but teachers will be able to see that in the $n$th line, where the sum is $n^3$, the first term of the consecutive odd numbers, whether $n$ is odd or even, is $n^2 - (n - 1)$, and the last is $n^2 + (n - 1)$.

The above pattern involving the sums of groups of consecutive odd numbers is only one of several interesting relationships connecting sums of integers – odd, even and both. Those which are quoted below offer scope for further investigation into inductive generalizations.

(1)

$$\begin{aligned} 1 &= 1^2 \\ 1 + 3 &= 2^2 \\ 1 + 3 + 5 &= 3^2 \\ 1 + 3 + 5 + 7 &= 4^2 \end{aligned}$$

and so on.

(2)

$$\begin{aligned} 2 &= 1^2 + 1 \\ 2 + 4 &= 2^2 + 2 \\ 2 + 4 + 6 &= 3^2 + 3 \\ 2 + 4 + 6 + 8 &= 4^2 + 4 \end{aligned}$$

and so on.

(3)

$$\begin{aligned} 2 &= 1^3 + 1 \\ 4 + 6 &= 2^3 + 2 \\ 8 + 10 + 12 &= 3^3 + 3 \\ 14 + 16 + 18 + 20 &= 4^3 + 4 \text{ etc.} \end{aligned}$$

(It will be seen that (2) is related directly to (1) and (3) is related to the example fully discussed in this section; but perhaps it is better that pupils should first examine the patterns independently.)

(4)

$$\begin{aligned} 1 &= 0^3 + 1^3 \\ 2 + 3 + 4 &= 1^3 + 2^3 \\ 5 + 6 + 7 + 8 + 9 &= 2^3 + 3^3 \\ 10 + 11 + 12 + 13 + 14 + 15 + 16 &= 3^3 + 4^3 \end{aligned}$$

and so on

(where there are successive groups of 1, 3, 5, 7, 9 . . . consecutive integers).

(5)

$$\begin{aligned} 1 &= 2(0^3 + 1^3) - 1 = 1^4 - 0^4 \\ 3 + 5 + 7 &= 2(1^3 + 2^3) - 3 = 2^4 - 1^4 \\ 9 + 11 + 13 + 15 + 17 &= 2(2^3 + 3^3) - 5 = 3^4 - 2^4 \\ 19 + 21 + 23 + 25 + 27 + 29 + 31 &= 2(3^3 + 4^3) - 7 = 4^4 - 3^4 \text{ etc.} \end{aligned}$$

**Example 7. Pythagorean triples**

(See also *Mathematics from Outdoors* and *Mathematical Pattern*)

These are sets of three whole numbers which can be the lengths of the sides of a right-angled triangle, e.g. 3, 4, 5; 8, 15, 17; etc.

It can be interesting to devise formulae or rules for finding such sets, and the teacher who wishes to make up questions on the Pythagorean relation which involve only integral square roots can also find such an activity valuable. Historically the problem goes back to the time of the Greeks and in succeeding centuries much ingenuity has been shown in continued investigation. In the following illustration only one simple approach is developed.

The pattern of square numbers increasing by successive odd numbers is well-known:

| | | | | | | | | | | | | |
|---|---|---|---|---|---|---|---|---|---|---|---|---|
| Square numbers | 1 | | 4 | | 9 | | 16 | | 25 | | 36 . . . | |
| Differences | | 3 | | 5 | | 7 | | 9 | | 11 | | . . . |

When there is a perfect square on the lower line also, this and the square of numbers which it is the difference represent the areas of the squares on the sides of a right-angled triangle.

Thus, 16 25
9 are three such square numbers, and the sides of the right-angled triangle are the square roots of these numbers, 3, 4, 5.

The next lower line perfect square is 25, this being the difference between 144 and 169. Hence the triangle with sides 5, 12 and 13 is right-angled.

The next odd perfect square in the difference 'line' (and of course all these numbers are odd) is 49, the difference between 576 and 625; so the triple 7, 24, 25 makes a right-angled triangle.

The question arises: is there an observable pattern which connects such numbers, from which subsequent triples can be predicted? If we examine the sets of three square numbers we have found so far, and write the square roots of the 'upper line' numbers just above them (fig. 12), then we have a clue to finding an answer to this question.

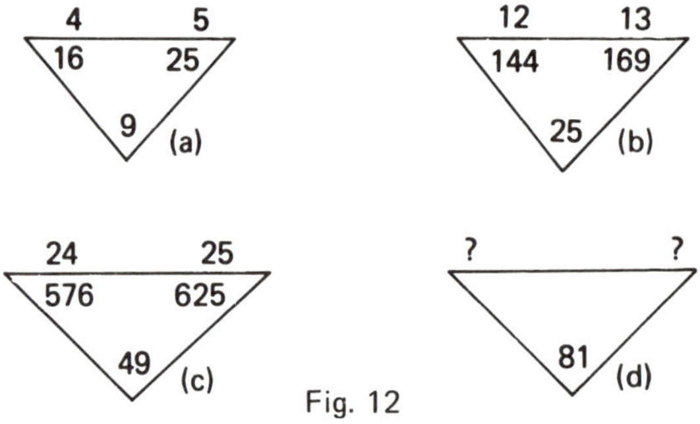


Fig. 12

Many pupils are quick to see, given this clue (and it probably will have to be given) that 9 is the sum of 4 and 5, 25 is the sum of 12 and 13, and 49 is the sum of 24 and 25.

The next 'lower line' odd perfect square is 81, so 'it looks as if' the square roots of the other two numbers should be . . .?

We may then check that these two numbers, with 9 (the square root of 81), form the sides of a right-angled triangle.

What numbers would be associated with 11 (the square root of 121)?

We have now completed stage one of our investigation, since we can find the 'next' numbers in the pattern.

How can we put into words an inductive generalization, which is stage two?

Such a generalization might run — in rather a cumbersome way — like this:

'The two numbers required have a sum which is the given odd perfect square and they are consecutive integers; so subtract 1 from that square number, and then halve it. This gives one of the numbers and the other will be 1 greater. The third number in the triple is the square root of the original odd perfect square.'

For example, take 1 from 49 and halve the result, giving 24; so 24 and 25 are two numbers in the triple, and 7 is the third.

(Teachers will be well aware of the keen interest many pupils find in testing number relationships and devices of this kind, even though they have no practical value. And the operations of squaring, adding, etc., which are involved are a self-imposed 'drill' which increases their facility with number in a far more effective way than through purposeless textbook examples.)

Stage three (only for teachers?) is the translation of the rule into symbols.

If a number is represented by $n$ then $2n$ will necessarily be an even number and $2n + 1$ an odd number.

A lower line odd square number will therefore be represented by $(2n + 1)^2$ which can also be written as $4n^2 + 4n + 1$.

Subtracting 1 from this leaves $4n^2 + 4n$, half of which is $2n^2 + 2n$.

The three numbers in the triple are therefore —

$$2n^2 + 2n;\ 2n^2 + 2n + 1;\ 2n + 1.$$

That these expressions do in fact give Pythagorean triples can be confirmed by putting $n$ equal to 1, 2, 3 . . ..

Or, in a more mathematically convincing way, by adding $(2n^2 + 2n)^2$ to $(2n + 1)^2$ and finding that the sum is in fact equal to $(2n^2 + 2n + 1)^2$.

An inevitable characteristic of all the whole number triples thus obtained is that two of the numbers are consecutive. There are many other Pythagorean triples where this is not the case.

It is interesting, as well as being mathematically very significant, that although the inductive pattern could only be discovered in the restricted range of whole numbers, the generalization allows *any* value for $n$ to be substituted in the symbolic expressions.

For example if

$$n = 1\tfrac{1}{2}$$

then

$$2n + 1 = 4 \text{ (which is not an odd number)}$$

$$2n^2 + 2n = 7\tfrac{1}{2}$$

$$2n^2 + 2n + 1 = 8\tfrac{1}{2}.$$

Thus the lengths 4, $7\frac{1}{2}$, $8\frac{1}{2}$ would make a right-angled triangle, the squares of these numbers being 16, $56\frac{1}{4}$ and $72\frac{1}{4}$ respectively:

$$16 + 56\tfrac{1}{4} = 72\tfrac{1}{4}.$$

If we double these three numbers, we obtain a whole number triple 8, 15, 17, whose squares are 64, 225 and 289;

$$64 + 225 = 289;$$

and this triple will give us a right-angled triangle, similar to the (4, $7\frac{1}{2}$, $8\frac{1}{2}$) triangle. We also obtain the triple (8, 15, 17), if we put $n = \frac{1}{3}$ in the symbolic expressions and multiply each of the resulting values by 9.

Again, if

$$n = 1\tfrac{1}{4},$$

then

$$2n + 1 = 3\tfrac{1}{2}$$

$$2n^2 + 2n = 5\tfrac{5}{8}$$

$$2n^2 + 2n + 1 = 6\tfrac{5}{8}.$$

Multiplying these by 8 gives the whole number triple 28, 45, 53.

Instead of using the symbolic generalization as above to extend the original type of triple, the verbal procedure (stage two) might be followed but we do not need to start with an odd perfect square.

For instance, take a square number which is even instead of odd, say 36.

Subtract 1, leaving 35; half of this is $17\frac{1}{2}$ and adding 1 gives $18\frac{1}{2}$.

The triple is therefore 6, $17\frac{1}{2}$; $18\frac{1}{2}$ or, in whole numbers, 12, 35, 37.

Again, suppose we choose $6\frac{1}{4}$, which is the square of $2\frac{1}{2}$.

Subtract 1, leaving $5\frac{1}{4}$; half of this is $2\frac{5}{8}$ and adding 1 gives $3\frac{5}{8}$.

Hence the triple is $2\frac{1}{2}$ , $2\frac{5}{8}$, $3\frac{5}{8}$ or, after multiplying by 8, is 20, 21, 29.

The curious-minded will want to see what is obtained when a square number less than 1 is selected, say $\frac{9}{16}$; or, using the symbolic rules, when $n$ is negative, e.g. $^{-}3$. And try $n = {}^{-}\frac{1}{2}$.

These are illustrations of the way a symbolism which has been invented to deal with a limited situation seems to take control and extends itself to cover ranges which previously had seemed to resist entry.

Another striking example of this is *index notation,* which by definition, starts by being meaningful only for whole numbers (e.g. $x^5$ is the shorthand for 'the product of five factors, each $x$') but which is ultimately extended to include indices which are fractional or negative, even though here one cannot use the original meaning. For example, $x^{5\frac{1}{2}}$ is NOT the product of $5\frac{1}{2}$ factors, each $x$.

**Example 8**

It is well-known that 4, 9, 16 . . . are called 'square' numbers because the number of dots representing them can be arranged in the form of squares.

For example

Fig. 13

The Greek mathematicians were very interested in such pictorial representations of numbers, and, lacking an efficient system of notation, discovered properties of triangular numbers, pentagonal numbers and still others, in addition to square numbers, by means of diagrams. They called them 'figurate' numbers. Another name for them is 'polygonal' numbers. Pupils often find a great deal of pleasure in investigating some of these numbers, and as well as the properties linked with their 'shapes' they can find many patterns and relationships in the numbers themselves. Many of these lend themselves to inductive reasoning at different levels, and a number of illustrations will now be given.

**Triangular numbers**

Fig. 14

Pupils quickly see that any triangular number is formed from the one before it by adding a longer 'base' to the triangle, containing one more point than the previous base (fig. 14).

If 1 is called the first triangular number, then the second has 2 points in each side of its triangle, the third has 3, the fourth has 4. The total numbers of points in turn are:

1, 3, 6, 10

We must add 5 for the next, 6 for the one after that, and so on.

If the number shapes are drawn on their 'sides' (fig. 15):

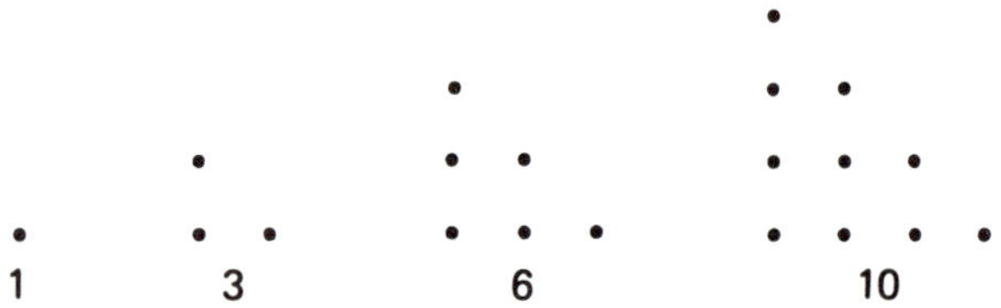


Fig. 15

and two consecutive triangles are put together, 'it looks as if' a square is formed each time (fig. 16).

Thus:

Fig. 16

This is obvious to us since we recognize

$$1 + 3 = 4; \quad 3 + 6 = 9; \quad 6 + 10 = 16; \quad \text{etc.}$$

as square numbers; but it is the 'shapes' which are of fundamental importance here. That is how the Greeks saw them, and how they will appear more significant to our pupils at first.

Inductively, therefore, we conclude that 'the sum of two consecutive triangular numbers is a square number'.

This verbal, inductive generalization is as far as we can go at the moment, but it will be seen later that it can be confirmed by the use of symbols.

It is clear that we can generate the sequence of triangular numbers by making sure that the difference between consecutive numbers in the sequence increases by 1 each time. The sequence begins with 1:

1 3 6 10 15 21 28 . . .

Differences 2 3 4 5 6 7 . . .

In this way we can find any triangular number, say the 10th or the 20th, but to do so one must know all the preceding ones. Should we wish to find the 100th triangular number this would be a very slow and tedious method. How could it be shortened?

Putting two identical triangular numbers together (omitting the ones's for the moment) we notice the 'shapes' which are formed (fig. 17):

```
                             o x x x x
              o x x x        o o x x x
o x x         o o x x        o o o x x
o o x         o o o x        o o o o x
```

Fig. 17

First we see that the shapes are rectangular; then that the number of points in the rectangles are, in turn;

2 x 3     3 x 4     4 x 5.

Inserting the first pair    o x

gives a recognizable pattern:

1 x 2     2 x 3     3 x 4     4 x 5

so that 'it looks as if' the next pair would be 5 x 6; and this can be checked. This is the first stage of inductive reasoning.

In this illustration it is not easy to go on to the second stage of expressing the pattern in a simple verbal way, and continued particular cases would probably have to be used as a bridge to the symbolic generalization.

Thus, from above, a pair of the *second* triangular numbers makes *2* x 3; a pair of the *third, 3* x 4; a pair of the *fourth, 4* x 5.

So (we are led to say) a pair of the *tenth* would be *10* x 11, and therefore the tenth triangular number is a half of this product, 55. (And this would be checked by counting on from 1, 3, 6, 10, . . .).

In the same way, a pair of the twentieth would be 20 x 21, giving the twentieth triangular number as 210.

We can write

$$T_4 = \frac{4 \times 5}{2} = 10$$

$$T_{10} = \frac{10 \times 11}{2} = 55$$

$$T_{20} = \frac{20 \times 21}{2} = 210$$

where $T_4$ is a shorthand for 'the fourth triangular number', and so on.

When pupils are ready for abstract symbols (and this will depend on previous experience with symbolic notation, the meaning of a bracket, etc.) the next step can be taken, where $T_n$ will be the shorthand for 'the $n$th triangular number', and so

$$T_n = \frac{n(n+1)}{2} \qquad \ldots (1)$$

An additional interpretation of this formula comes from looking again at the 'pictures' of triangular numbers (p. 31).

Each of these numbers is derived from the previous one by increasing the side, or the bottom row, by 1.

That is, the 4th triangular number is $1 + 2 + 3 + 4$; the 9th is $1 + 2 + 3 + \ldots + 8 + 9$; and in general the $n$th is

$$1 + 2 + \ldots + n$$

It can therefore be seen from (1) above that

$$1 + 2 + 3 + \ldots + n = \frac{n(n+1)}{2}$$

i.e. that the sum of $n$ consecutive integers starting with 1 is

$$\frac{n(n+1)}{2} \text{ or } \frac{1}{2}n(n+1).$$

We recognize that many of our pupils may not reach all the stages in the development outlined above. How much of the work they do complete will be governed by the judgement of the teacher. The step we now describe could well be beyond the powers of any of those pupils we are concerned with, since further abstractions are involved, as well as algebraic manipulations which would in any case come late in the course. It is included, however, largely to complete for teachers this particular sector of discussion and it is possible that it will be appropriate for some pupils.

We found inductively that the 'sum of two consecutive triangular numbers is a square number'. That is, 'it looks as if' this is always the case. We can now show *deductively* that it is so, accepting that the rule or formula for the $n$th triangular number is

$$\frac{n(n+1)}{2}.$$

Since, say, the twenty-third triangular number is found by replacing $n$ in this formula by 23, the $(n-1)$th triangular number – the one immediately before the $n$th – is expressed by replacing $n$ by $(n-1)$.

This gives $T_{n-1}$ to be

$$\frac{(n-1) \times (n-1+1)}{2} \quad \text{or} \quad \frac{(n-1)n}{2}.$$

The sum of the consecutive triangular numbers $T_{n-1}$ and $T_n$ is therefore

$$T_{n-1} + T_n = \frac{n(n-1)}{2} + \frac{n(n+1)}{2}$$

$$= \frac{n^2 - n}{2} + \frac{n^2 + n}{2}$$

$$= \frac{1}{2}n^2 - \frac{1}{2}n + \frac{1}{2}n^2 + \frac{1}{2}n$$

$$= n^2$$

which is, of course, a square number.

Thus we have now firmly established the *inductive* conclusion by deduction.

**Example 9. Intersections of straight lines on a plane**

There are two interesting investigations: firstly, the number of points of intersection made by 1, 2, 3, 4, . . . lines; secondly, the number of 'spaces' or regions into which the plane is divided by these lines, supposing the lines to be of indefinite length (fig. 18).

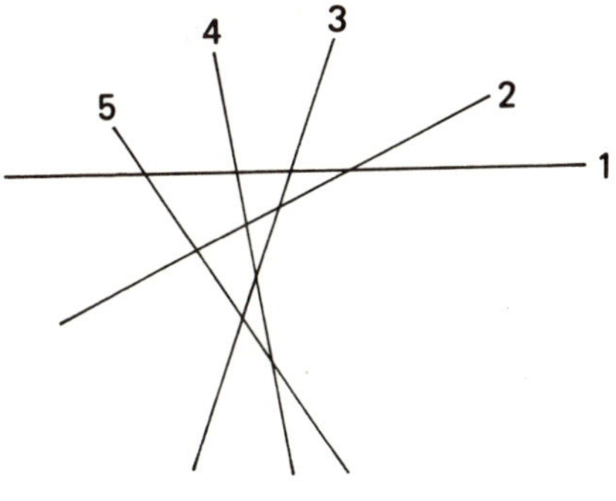


Fig. 18

An inspection of a drawing, such as fig. 18, gives the numbers in the table below:

| | | | | | |
|---|---|---|---|---|---|
| Number of lines | 1 | 2 | 3 | 4 | 5 |
| Number of point intersections | 0 | 1 | 3 | 6 | 10 |
| Number of spaces | 2 | 4 | 7 | 11 | 16 |

The pattern of point intersections seems to be the triangular numbers; is this confirmed when six lines intersect?

The pattern of spaces is not a familiar one, but teachers will possibly find some pupils who perceive that each of these numbers is one greater than a triangular number (assuming, of course, that this series has become familiar).

Another relation which pupils have pointed out is that the sum of the numbers of lines and points is itself always a triangular number, and is one less than the number of spaces.

If and when the stage of symbolic representation is reached, it will be seen that the rule for the *n*th triangular number, namely, $\frac{1}{2}n(n + 1)$, gives the number of point intersections one place ahead of what it really is. For example the fourth triangular number is 10, but when the number of lines is four, the number of points is six, the *third* triangular number.

This is because the first number in our sequence is 0 and the triangular numbers start one place later; in other words, $n$ in the formula should be replaced by $(n - 1)$, giving $\frac{1}{2}(n - 1) \times n$ or $\frac{1}{2}n(n - 1)$ as the inductive rule for the pattern of point intersections.

For the number of spaces, however, the numbers are each one more than the triangular numbers in their proper position; the formula for the *n*th number in this pattern will therefore be $\frac{1}{2}n(n + 1) + 1$ or $\frac{1}{2}(n^2 + n + 2)$.

It is not so surprising that the triangular numbers enter into the pattern of point intersections if we consider that the second line we draw will cut the first in *one* point; the third line will cut the other two in *two* points; the fourth line will cut the other three in *three* points, and so on. The total number of points is $1 + 2 + 3 + \ldots$ which as we know is a pattern which generates the set of triangular numbers.

It will be found that the same pattern of numbers is obtained when straight lines are drawn irregularly across a circle, provided that each line crosses each of the others (fig. 19). From this fact we infer that

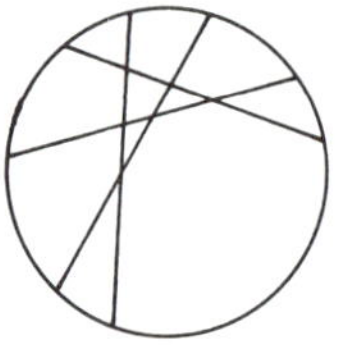

Fig. 19

the same relationship holds, this time between the number of lines and the maximum number of regions made in the circle. (The word 'maximum' is included to avoid special cases obtained by making several lines pass through the same point.)

On the other hand, if regions are made in a circle by joining 2, 3, 4, . . . points on the circumference in all possible ways, the pattern relating *points* and regions at first seems to be clear, but then breaks down. (This illustration was referred to in chapter 1.) There is, in fact, a definite pattern for which a symbolic formula can be obtained, but not inductively, and it is not a simple one.

Teachers may be interested to extend these investigations by trying to find the pattern for the number of three-dimensional spaces made by 1, 2, 3, 4, . . . *planes.* The problem very quickly becomes difficult to tackle either diagrammatically or in imagination; reference to chapter 3 of Polya's *Mathematics and Plausible Reasoning* (O.U.P.) will probably be necessary to satisfy curiosity as well as to see how inductive reasoning gives the clues to the solution. (In case readers think they have found a very simple pattern, it should be said that with four intersecting planes the number of spaces is *not* 16.)

**Example 10**

Work out the values of

$$1^3;\ 1^3 + 2^3;\ 1^3 + 2^3 + 3^3;\ 1^3 + 2^3 + 3^3 + 4^3.$$

What special kind of numbers are obtained? What does it appear that the 'next' relation might be?

'It looks as if' the sum of consecutive cube numbers starting with $1^3$ is each time a square number.

Can the inductive generalization be taken a stage further?

These square numbers when written as $1^2, 3^2, 6^2, 10^2, \ldots$ reveal themselves as the squares of triangular numbers. Does the sum of the first five cubic numbers give the square of the next triangular number?

The verbal generalization therefore becomes: The sum of a number of consecutive cubic numbers, starting with $1^3$, is equal to the square of that 'order' of triangular numbers.

Now we have previously found that the $n$th triangular number can be written as

$$\frac{n(n+1)}{2},$$

so the simple symbolic generalization of the statement is:

$$1^3 + 2^3 + 3^3 + \ldots + n^3 = \left[\frac{n(n+1)}{2}\right]^2 \text{ or } \frac{1}{4}n^2(n+1)^2.$$

In example 8 it was shown that $1 + 2 + 3 \ldots + n = \dfrac{n(n+2)}{2}$.

We may now state that:

*The sum of cubes of consecutive integers, starting with* $1^3$ *is equal to the square of the sum of the same consecutive integers,*

since $\frac{1}{4}n^2(n+1)^2$ is the square of $\dfrac{n(n+1)}{2}$.

**Example 11**

The relation

$$3^2 + 4^2 = 5^2$$

in which three consecutive integers are squared is well-known.

It is not so well-known that

$$10^2 + 11^2 + 12^2 = 13^2 + 14^2$$

in which five consecutive numbers are squared.

If there is a pattern in such sequences, the next should contain seven such numbers, four on the left and three on the right; in fact the relation is

$$21^2 + 22^2 + 23^2 + 24^2 = 25^2 + 26^2 + 27^2.$$

Is there any clue which will enable us to say 'It looks as if the next line should be . . .'?

The clue lies in the recognition of triangular numbers, which turn up in number patterns with surprising frequency. The induction will be left to the reader.

**Example 12. Polygonal numbers**

Other 'figurate' numbers besides the triangular and the square numbers have an intrinsic interest, but the relationships between them also illustrate vividly the power of inductive reasoning so we shall now investigate them in some detail.

It is instructive to follow up triangular and square numbers with the idea of the possibility of 'pentagonal' numbers, and to attempt to draw their shapes and speculate on how many points they would contain.

Conventionally the first has one point, and the second will clearly have five, but how many are there in the third? After pupils and teacher (and his colleagues?) have drawn pentagonal number shapes and have argued about them it will be seen that there appear to be several possibilities, and before any conclusion is drawn perhaps the following development might be considered.

The whole numbers 1, 2, 3, 4, . . . may permissibly be described as 'linear' numbers since they can be represented thus:

•   • •   • • •   • • • •   etc

Fig. 20

The following table of numbers can be drawn up:

| Linear | Triangular | Square | Pentagonal |
| --- | --- | --- | --- |
| 1 | 1 | 1 | 1 |
| 2 | 3 | 4 | 5 |
| 3 | 6 | 9 | |
| 4 | 10 | 16 | |
| 5 | 15 | 25 | |
| 6 | 21 | 36 | |
| ⋮ | ⋮ | ⋮ | |

How many patterns can now be discovered?

(Pupils can be relied on to find some.) What inductions can be made?

For example,

(a) The second row numbers increase by ones, the third row (as far as it goes) by threes, the fourth by sixes, the fifth by tens, the sixth by fifteens,

i.e. by 1, 3, 6, 10, 15 successively.

But these are the triangular numbers, so 'it looks as if' the seventh row should increase by 21's; and this is so, since the first three numbers in this row are 7, 28, 49.

(b) Since the third row is increasing by 3 each time, will the next pentagonal number be 12?

This is an induction which needs to be checked. Do any of the shapes drawn for pentagonal numbers seem to confirm it?

From the patterns suggested by the fourth, fifth and sixth rows it appears that the next pentagonal numbers should be 22, 35, 51. Do the diagrams confirm these numbers? And we can find any corroborative evidence? (The next section brings it to light.)

(c) Look next at the successive differences in the columns.

| Linear | Triangular | Square | Pentagonal |
| --- | --- | --- | --- |
| 1 | 1 | 1 | 1 |
| (1) | (2) | (3) | (4) |
| 2 | 3 | 4 | 5 |
| (1) | (3) | (5) | |
| 3 | 6 | 9 | |
| (1) | (4) | (7) | |
| 4 | 10 | 16 | |
| (1) | (5) | (9) | |
| 5 | 15 | 25 | |

These differences (ringed in the table) in turn *increase* by 0 (being all ones); by 1 each time in the second column; by 2 each time in the third column.

May we make the induction that in the column of pentagonal numbers, they would increase by 3? If so, the next difference after the 4 would be 7, making the third pentagonal number 12 – which agrees with our previous surmise. And succeeding differences would be 10 and 13, giving the fourth and fifth pentagonal numbers as 22 and 35, which are again in agreement. The induction seems to be correct.

But if the 'differences' (the ringed numbers) are examined horizontally, still another induction can be made, providing further evidence for the sizes of the pentagonal numbers. (This induction will be left to the reader.)

If the 'shapes' of these numbers do not seem to fit into a pattern, perhaps the following idea will help:

Each triangular number is obtained from the previous one by *adding* points which preserve the characteristic shape, and have respectively 2, 3, 4, . . . points in each external side.

For example, to change from  to 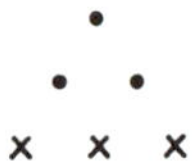

a base of three is added and the side of the triangle changes in length from two points to three points.

Again, changing from  to 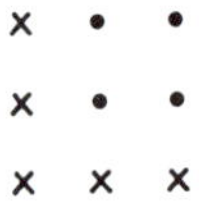

involves adding the L-shape which increases the points in each side from two to three.

The first step in changing the second pentagonal number to the third should perhaps therefore be to add points to make three in each of two sides (at first) as shown:

Fig. 21

The pentagonal shape must then be built up so that each of the five sides contains three points. This suggests the shape:

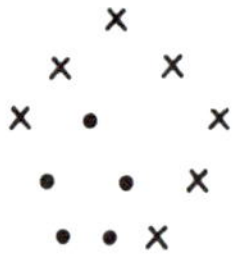

Fig. 22

It is true that this does not give as satisfying a picture as with triangular and square numbers, but it is a logical procedure and it certainly contains 12 points. If the fourth pentagonal number is built up it will be found to have 22 points, as the patterns in the table suggest.

(d) The table can now be extended indefinitely as a consequence of the several inductions, which all corroborate one another.

| Linear ($L$) | Triangular ($T$) | Square ($S$) | Pentagonal ($P$) | Hexagonal ($H$) | Heptagonal ($Ht$) |
|---|---|---|---|---|---|
| 1 | 1 | 1 | 1 | 1 | 1 |
| 2 | 3 | 4 | 5 | 6 | 7 |
| 3 | 6 | 9 | 12 | 15 | 18 |
| 4 | 10 | 16 | 22 | 28 | 34 |
| 5 | 15 | 25 | 35 | 45 | 55 |
| ⋮ | ⋮ | ⋮ | ⋮ | ⋮ | ⋮ |

It is now possible to discover several other relationship patterns, and again it may be said that pupils often look for them excitedly when perhaps one or two hints have been given.

For conciseness in the writing at this point, these inductions will be stated at once in symbols. It must be understood, however, that with pupils the normal stages would be followed. That is, first of all, some of the numbers in the table are seen to be related; then a *pattern* of relationships is noticed – implying an inductive generalization; and next a verbal description of this generalization is made (for example 'a triangular number in one row added to the square number in the next row seems always to equal the pentagonal number in this row').

Such a development must not be regarded as incomplete or unsatisfactory if this is as far as some pupils can go. Indeed, if they achieve this verbal generalization stage, their mathematical thinking has been quite significant, and perhaps only a few will get so far.

Here, however, is a list of such generalizations expressed symbolically.

(1) $T_{n-1} + S_n = P_n$
(2) $T_{n-1} + P_n = H_n$
(3) $T_{n-1} + H_n = Ht_n$

where $T, S, P, H, Ht$ stand for triangular, square, pentagonal, hexagonal, heptagonal numbers respectively.

(and so on – this phrase itself being another inductive generalization).

(4) $L + P \; = T + S$
(5) $T + H \; = S + P$
(6) $S + Ht = P + H$

(in each horizontal row)

And so on.

(7) $T_{n-1} + T_{n+1} = 2T_n + 1$
(8) $S_{n-1} + S_{n+1} = 2S_n + 2$
(9) $P_{n-1} + P_{n+1} = 2P_n + 3$

(Now make an induction about another possible relation, and test it.)

(10) $T_{n-2} + T_{n+1} = T_{n-1} + T_n + 2$
(11) $S_{n-2} + S_{n+1} = S_{n-1} + S_n + 4$
(12) $P_{n-2} + P_{n+1} = P_{n-1} + P_n + ?$

And so on.

It may be remarked that (7) and (10); (8) and (11); (9) and (12) are not independent pairs of relations.

The inductive relations (1), (2) and (3) can be used to find formulae for the general terms of the respective columns, namely the pentagonal, hexagonal, heptagonal, . . . numbers. These developments, are of course, rapidly moving out of the range of the pupils we are concerned with.

Before relation (1) is analysed symbolically, however, it is interesting to see how it is illustrated pictorially. The assertion is that a triangular number added to the 'next' square number gives the 'next' pentagonal number, e.g. 3 + 9 = 12; 6 + 16 = 22; 10 + 25 = 35; etc.

The first two give the pictures:

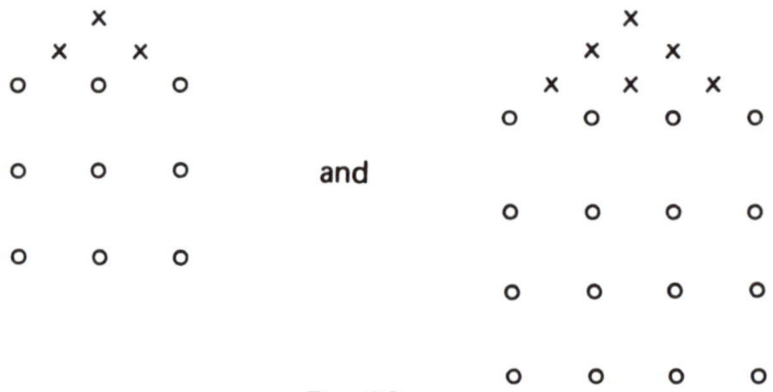


Fig. 23

which are perhaps simpler diagrammatic patterns than those previously illustrated.

Returning to the symbolic expression of this relation, the $n$th triangular number has been found to be $\frac{1}{2}n(n+1)$ so that the $(n-1)$th such number is $\frac{1}{2}(n-1)n$.

The $n$th square number is $n^2$ (for the 5th is $5^2$, the 9th is $9^2$, etc.)

Formula (1) on p. 41 gives

$$P_n = T_{n-1} + S_n$$
$$= \frac{1}{2}n(n-1) + n^2$$
$$= \frac{3}{2}n^2 - \frac{1}{2}n$$

i.e.
$$P_n = \frac{1}{2}n(3n-1).$$

This may be tested for $n = 5$, say, giving the 5th pentagonal number as $\frac{1}{2} \times 5 \times 14$, i.e. 35 which agrees with the table.

The rule for the hexagonal numbers comes from formula (2) on p. 41 which is

$$H_n = T_{n-1} + P_n$$
$$= \frac{1}{2}n(n-1) + \frac{1}{2}n(3n-1)$$
$$= \frac{1}{2}n^2 - \frac{1}{2}n + \frac{3}{2}n^2 - \frac{1}{2}n$$
$$= 2n^2 - n$$

i.e.
$$H_n = n(2n-1).$$

(Putting $n = 5$ gives $H_5 = 5 \times 9 = 45$
and $n = 6$ gives $H_6 = 6 \times 11 = 66$
as the table states.)

Similarly from (3) the formula for the $n$th heptagonal number is found to be $\frac{1}{2}n(5n-3)$.

Instead of using relations (1) (2) and (3) to derive these formulae, relations (4), (5) and (6) may be used instead. For example, relation (4) is true for each horizontal row, so it applies to the $n$th row and so to the $n$th terms.

That is,

$$L_n + P_n = T_n + S_n.$$

The $n$th linear number is $n$, and the formulae for $T_n$ and $S_n$ are known, so:

$$n + P_n = \frac{1}{2}n(n+1) + n^2$$

$$\therefore \qquad P_n = \frac{1}{2}n^2 + \frac{1}{2}n + n^2 - n$$

$$= \frac{3}{2}n^2 - \frac{1}{2}n$$

$$\therefore \qquad P_n = \frac{1}{2}n(3n - 1) \text{ as before.}$$

Similarly, from (5)

$$T_n + H_n = S_n + P_n$$

$$\therefore \qquad \frac{1}{2}n(n + 1) + H_n = n^2 + \frac{1}{2}n(3n - 1)$$

$$\therefore \qquad H_n = n^2 + \frac{3}{2}n^2 - \frac{1}{2}n - \frac{1}{2}n^2 - \frac{1}{2}n$$

$$= 2n^2 - n$$

$$\therefore \qquad H_n = n(2n - 1).$$

In the same way, relation (7) leads again to the formula

$$Ht_n = \frac{1}{2}n(5n - 3).$$

The manner of derivation of these formulae illustrates how deductive reasoning may be applied to inductively obtained generalizations to produce new results.

The formulae derived in these two ways (and there are still other ways of obtaining them) yield further inductions when written as follows:

$$L_n = n$$

$$T_n = \frac{1}{2}n^2 + \frac{1}{2}n$$

$$S_n = n^2 + 0 \,.\, n$$

$$P_n = 1\frac{1}{2}n^2 - \frac{1}{2}n$$

$$H_n = 2n^2 - n$$

$$Ht_n = 2\frac{1}{2}n^2 - 1\frac{1}{2}n.$$

The coefficients of the $n^2$ term are successively 0, $\frac{1}{2}$, 1, $1\frac{1}{2}$, 2, $2\frac{1}{2}$, so it certainly looks as if' the one for the octagonal number will be 3.

The coefficients of the term in *n* are, respectively,

$$1 \quad \frac{1}{2} \quad 0 \quad \frac{^-1}{2} \quad {}^-1 \quad {}^-1\frac{1}{2}$$

What does it appear that the next will be?

These are inductive conclusions. What do they suggest that the formula for the *n*th octagonal number should be? If in this surmised formula *n* is successively put equal to 1, 2, 3, 4, . . . the calculated numbers are

$$1, 8, 21, 40, \ldots$$

Now if the table on p. 41 is extended to the next column, either by the horizontal pattern or by the pattern of vertical 'differences', it will be seen that the same numbers are obtained, so the predictions of inductive reasoning are checked.

It is recognized that most of the above manipulations with symbols lie outside the range of probably nearly all 'our' pupils. They have been included because they illustrate the value of inductive reasoning in mathematics not only in purely numerical contexts, but also where symbolic generalizations are concerned. They also show how number patterns which pupils can discover and enjoy at a simple level can lead to relationships at progressively higher levels – to mathematics which is, as it were, self-generating.

To summarize, the purpose of this chapter has been to emphasize the importance of generalization, the use of inductive reasoning whereby generalizations can be made, and – only to the extent that individual pupils can profit – the value of symbols in translating the generalizations from the 'English' language into concise 'mathematical' language and the function of manipulative techniques in using the symbolic rules or formulae. Although the contents of this chapter will no doubt provide material which is interesting and appropriate for pupils at different stages and levels of ability, it is hoped that the examples will help the teacher to be on the look-out for many other opportunities for generalization in mathematics, and will show how pupils can be guided to the discovery of relationships by inductive reasoning. In the next chapter illustrations will be given of the use of induction in formulating some of the standard algebraic techniques, and once again it should be appreciated that these are typical of many other possible applications of the method.

# 3 The understanding and use of formulae

In chapter 1 we emphasized the importance of the formula as the working tool of the engineer, the technician and the craftsman. We also stressed that a formula is the symbolic expression of a relationship, which in simple cases will first have been reached and stated in words; it expresses the connection between quantities which the technicians and scientists have to measure. Although these quantities change in value under different conditions, the way in which they are related remains the same, and the formula is the generalization which covers all the particular situations for which its application is valid.

Sometimes formulae are derived by way of theoretical arguments and logical inference; in other cases they are 'empirical' i.e. they are the result of experiments and the formulation of working rules. It is not necessary for the engineer, say, to know how his formula is obtained, but he must know what it means and how to use it.

For example, a wire stretched between two supports has to carry a weight, what is the pull on the supports?

The formula which applies to the situation is

$$P = \frac{LM}{2S}$$

where $L$ is the length of the wire and $S$ the sag (both in centimetres), $M$ is the mass of the load and $P$ the pull (both in kilogrammes).

Again the weight ($W$) which can be supported by a floorboard of length $L$, breadth $b$ and thickness $t$ (all in centimetres) might be, in kilogrammes:

$$W = \frac{0{\cdot}6bt^2}{L}$$

the numerical factor varying with the type of timber.

The latter part of chapter 1 illustrated briefly how the understanding of formulae and the way they are used involve various manipulative techniques which are usually accepted as a major part of an algebra course. In this chapter we now look at some of the techniques and processes in greater detail, with suggestions on how pupils might be led, through both inductive and deductive reasoning, to discover and formulate them in a meaningful way, rather than merely to be told what the rules are, as often happens.

In the first place, how should pupils become acquainted with formulae? A number of simple formulae grow out of mensuration rules which the wise teacher helps his pupils to evolve out of practical experience. These are discovered over several years of mathematical study, starting in the primary school and continuing through the secondary course. In nearly all cases it is a sound principle to have these rules expressed in words and to avoid the symbolic forms until they are thoroughly understood and have been used with understanding in the verbal way for calculation purposes. In fact this practice is followed in many handbooks and instruction books for engineers, craftsmen and the do-it yourself handyman. An example taken from motorcar engineering concerns the calculation of the capacity of a car engine. The rule quoted in the handbook is:

> 'Square the bore in cm, multiply by 0·7854 and multiply the result by the stroke in cm and the number of cylinders. The result is in c.c.'

Another rule, for calculating the number of engine revolutions per minute at a given speed (in kilometres per hour) uses a layout which is 'mathematical' but still uses words rather than symbols:

$$\text{r.p.m.} = \frac{530 \times \text{speed} \times \text{gear ratio}}{\text{wheel diameter in centimetres (including tyre.)}}$$

Examples of verbal generalizations in use in the classroom are:

the area of a rectangle is the number of squares in a row multiplied by the number of rows; or (a later version) the length multiplied by the breadth;
the area of a triangle is half the area of the rectangle it belongs to;
the circumference of a circle is $\pi$ diameters;
the area of a circle is $\pi$ squares on the radius;
the volume of a cuboid or cylinder is the area of the base multiplied by the perpendicular height.

The time comes, but it should not be hurried, when it is appropriate to suggest – and sometimes the suggestion comes spontaneously from the pupils themselves in written work – that it is quicker and simpler to write some of these words in a shorthand form. Thus, *C* can obviously be used to represent the circumference of a circle and $D$ its diameter; the rule then becomes

$$C = \pi \times D.$$

Similar condensations can be made for other rules which pupils need to use.

It will thus be appreciated – and this is a major point –

*that the letters used in formulae are replacements for words and phrases.*

Many of the older textbooks say in effect that in algebra letters are used instead of numbers, or that letters 'stand for' numbers. This assertion, made in the context of symbols which have no physical significance, is probably the root cause of the puzzlement many pupils experience in studying algebra. The idea of algebra as 'generalized arithmetic' if this is the sole approach, as it often is, can be quite misleading, and tends to cause confusion in the pupil's mind. In the words of T. P. Nunn, 'he cannot easily see how $x$ can mean a particular number without meaning either *this* particular number or *that* one'. But a number which is substituted in a formula is readily accepted as a replacement for a measurement in a particular situation. The shorthand letter is a description of this measurement (a diameter, a temperature, etc.), and in different situations it is only reasonable that different numbers should be appropriate.

As we mentioned in chapter 1, there are conventions which make for further conciseness in changing a verbal rule into a symbolic formula, and these will normally be introduced when the teacher feels the time is appropriate. Little harm is done by delaying somewhat the introduction of a definition, technical term or process. Introducing any of these prematurely can lead only to misunderstanding. Thus the rule expressed in words for finding the area of a circle will first be written in the shorthand form

$$A = \pi \times (r \times r)$$

Two further modifications have to be made before it becomes, more conventionally,

$$A = \pi r^2$$

Teachers will be well aware of the confusion which exists in the minds of many pupils between the expressions $2r$ and $r^2$. It is through our insisting on real situations (pictorial or tangible) with verbal descriptions to preface or illustrate the symbolic expressions that the distinction is more likely to become clear.

For example, the distance around a square is 'the lengths of four sides added together', or alternatively, 'four times the length of one side'.

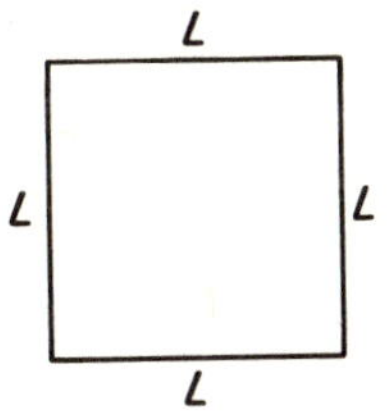


Fig. 24

(It is appreciated that even to achieve statements such as these demands patient and painstaking effort from the teacher of 'below average' pupils.)

In symbols, the above statements reveal the identity of

$$L + L + L + L \quad \text{and} \quad 4 \times L \quad \text{or} \quad 4L.$$

The area of the square (the length multiplied by the breadth) shortens to

$$A = L \times L \quad or \quad A = L^2$$

By using concrete illustrations and asking our pupils to make statements about them in words, rather than expecting them to juggle with meaningless letters, we are helping our pupils to make correct statements as a matter of course. At a later stage they can proceed to more abstract and mechanical work with a confidence born of this earlier success.

If, and when, we need to concentrate on the 'collection of like terms', we can use the perimeters of shapes. Examples would first be worked out using actual lengths; the method of calculating the perimeter would next be described in words, and the transition to a shorthand formula using letters would follow.

For example, (following numerical calculations which are omitted here)

(1) Distance around fig. 25(a) =

$W + L + L + W + \ldots + L$ (in order)

= 4 widths + 8 lengths

$= 4W + 8L.$

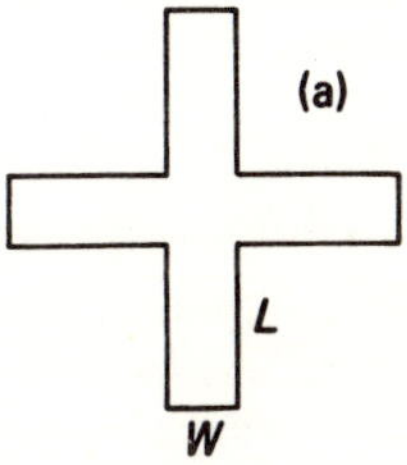


Fig. 25(a)

(2) Distance around fig. 25(b)

$= - \ - - - - - - - -$

$= 4W + 6L$

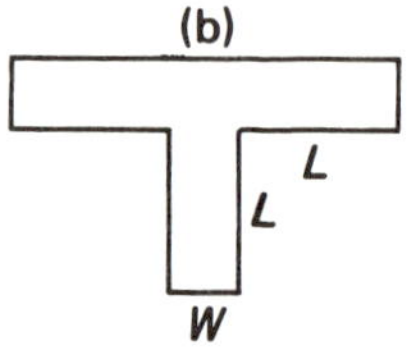


Fig. 25(b).

Structural apparatus of many kinds can be put to use, with pupils making their own shapes. Alternatively we can give pupils lengths of dowelling rod, some of one fixed length and some of another, and ask them to make shapes (or even straight lines) with them. The total length of dowelling would then be represented in symbols by expressions like $5L + 7S$, where $L$ stands for the length of a long piece and $S$ for that of a short one.

Returning to the matter of how pupils can gain familiarity with formulae, we stress the need for a great deal of practice both in devising the symbolic shorthand for verbal rules and in explaining in words what the symbols in particular formulae mean and how they are to be operated on. To be able to translate concisely a lengthy statement in the 'English language' into 'mathematical language', and conversely, gives a mastery over symbols and an understanding of them which is basic to algebra, and hence to other branches of mathematics. As mentioned earlier, it is very true that most pupils find it extremely difficult to make statements in a precise way about the relationships they perceive so that they are indeed 'mathematical', but they can be helped considerably by being given opportunities for handling material and equipment and by actively performing operations with it. Frequent practice in this kind of 'translation' should be given throughout an algebra course, and indeed whenever symbols are used in mathematics. Many algebra texts contain suitable examples for pupils; a mathematics department library should have a wide range of books from which teachers can make appropriate selections. A few typical examples are given here. Again it is helpful to start with a numerical illustration. By considering how such an example behaves it is easier to put into words the general rule which yields the symbolic formula. Sometimes a pattern would lead to the rule inductively.

(1) A number of seconds elapse between our seeing a flash of lightning and hearing the thunder. The interval in seconds ($t$) divided by 8 gives the distance away of the flash in kilometres ($D$).

$$D = \frac{t}{8}$$

(2) There are 40 litres of water in a bath. Water runs in at the rate of 15 litres a minute. What is the formula for the number of litres in the bath after $t$ minutes?

$$(L = 40 + 15t)$$

(3) A framework is made out of 12 pieces of metal (fig. 26) and canvas is stretched over all the faces. What are the formulae for

(i) the total length of metal used? $(L = 8a + 4b)$

(ii) the area of the canvas used? $(A = 2a^2 + 4ab)$

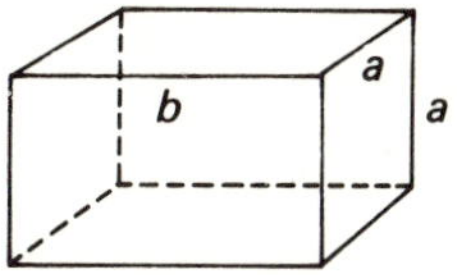


Fig. 26

(4) A piece of elastic a metre long is fastened to a hook so that it hangs vertically. Every successive weight of 10 grammes which it supports extends it a further centimetre. What is the formula connecting $L$, its stretched length in cm, with $W$ the supported weight in grammes?

$$L = 100 + \frac{W}{10} .$$

(5) In Rugby Union football an unconverted try counts 4 points and a goal (a converted try) counts 6 points. In a match $T$ tries are scored of which $G$ are converted into goals. What is the score? Different lines of reasoning lead to

$$S = 4T + 2G$$

and $$S = 4(T - G) + 6G.$$

At a later stage a comparison of these formulae will bring to light a property or rule about brackets.

## Indices

A detailed development is not necessary here, but we make certain observations.

1. Familiarity with index notation using numbers should come before we attempt to genralize using letters. E.g. powers of 10 for the concise expression of very large numbers (astronomical distances, etc.); powers of 2, as related to base two numbers, or to the legend of the origin of

the game of chess and similar stories; and more generally the representation of numbers in various bases.

It is very important that pupils should understand and use index notation long before operations are considered. The same principle applies to the use of other representations of number, such as whole numbers and different kinds of fractions. Very often the difficulties pupils find in 'the four rules' can be traced back to lack of familiarity with notation.

2. Very careful explanation of the notation is essential, so that there is no likelihood of, say, $2^5$, which may be described as a shorthand for 'five two's all multiplied together', being written or thought of as $2 \times 5$. The phrase sometimes heard '2 multiplied by itself 5 times' is incorrect, since 2 multiplied by itself once is $2 \times 2$ or $2^2$. A useful form of words says that $2^5$ is *one* multiplied by two five times or $1 \times 2 \times 2 \times 2 \times 2 \times 2$. Not only is this mathematically correct, but it incorporates both the 2 and the index 5 in the same statement. Furthermore when $2^0$ is being considered, this statement reduces to '$2^0$ is *one* multiplied by two no times' and this result is plausibly 1. Hence $2^0 = 1$ becomes a credible statement.

When the terms 'factor' and 'product' are properly understood (and it must be admitted that generally they are not used frequently enough for pupils to become familiar with them) an exact description of $2^5$ is 'the product of five factors, each 2'.

Whatever the terminology, much practice must be given in writing both numbers and expressions which use letters, in equivalent longhand and shorthand forms.

3. In considering *operations* on numbers written in index form, we must take care to distinguish between those operations which are valid and those which are not.

For example, the sum, or difference of $3^4$ and $3^2$ (same base, different powers) cannot be expressed as whole number powers of 3 or any other base. (Care must be taken in the choice of examples; the difference of consecutive powers of 2 is itself a power of 2.) Nor can $3^4 \pm 2^4$ (same powers, different bases) be so expressed.

Other counter examples, with different powers, different bases, should be taken as illustrations.

4. In obtaining the rules for multiplication and division in index notation an appeal to inductive reasoning is in order, but care should be taken not to force a generalization from too few examples.

A typical illustration might be $2^3 \times 2^5$. Changing from 'shorthand' to 'longhand', this is the same as $2 \times 2 \times 2 \times 2 \times 2 \times 2 \times 2 \times 2$ for which the shorthand notation is $2^8$.

After a number of examples have been done in this way the equations should be collected together in a list which can be examined by the pupils.

Thus:

$$2^3 \times 2^5 = 2^8$$
$$3^2 \times 3^4 = 3^6$$
$$10^4 \times 10^5 = 10^9$$
$$5^3 \times 5^2 = 5^5$$
------ -- ------ = ------
------ -- ------ = ------

Discussion quickly leads to the conclusion that 'it looks as if . . . etc'.

Later on, the zero index and then negative indices will be introduced. For these, the building up of patterns from which inductions can be made gives convincing evidence of interpretations which seem reasonable, although the common-sense of the original shorthand has disappeared. (For $5^0$ is not the product of zero factors, each 5; nor is there any meaning in the statement 'the product of 3 negative factors, each 5' as a definition of $5^{-3}$.)

Write, for example:

$$10^5 = 100\,000$$
$$10^4 = 10\,000$$
$$10^3 = 1000$$
$$10^2 = 100$$
$$10^1 = 10$$
------ -- -------- ----

On the left side, the pattern shows the index decreasing by 1 in successive lines, so the next will be $10^0$.

On the right side, each line is divided by 10 to obtain the succeeding line, so the next will be 1.

So it looks as if $10^0$ should be represented by 1.

Similar chains of numbers lead to the interpretation of $2^0$, $3^0$, $5^0$, etc. each as 1. This is of course in agreement with one of the word forms given on p. 52.

Corroborative evidence is supplied by the traditional text-book 'derivation' where, say, $10^3$ is divided by $10^3$ to give 1, and also to give $10^0$ by applying the index rule for division. Expressing these thoughts symbolically:

since $\dfrac{10^3}{10^3} = 1$

and $10^3 \div 10^3 = 10^0$ (by the division rule for indices)

then $10^0 = 1$.

Another way to provide supporting evidence for pupils who have had plenty of experience of graphs is to draw on a large scale the curve which passes through $2^4$, $2^3$, $2^2$, $2^1$ (from horizontal axis numbers 4, 3, 2, 1 respectively) (fig. 27). There is an unreasonable 'kink' if one extends to it go through the origin, where it might possibly have been

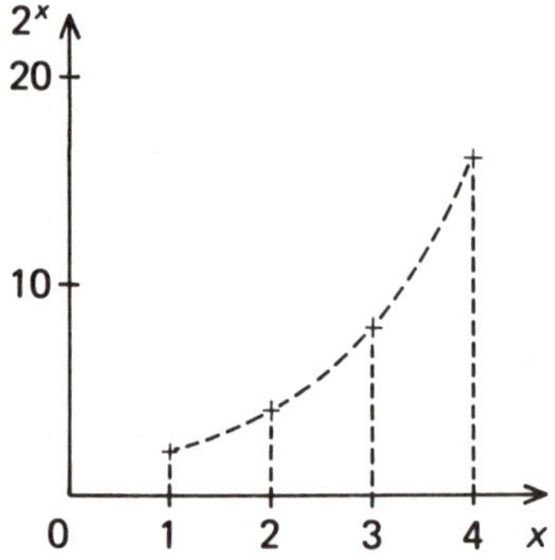


Fig. 27

expected to go. (Interpolation on this curve, incidentally, also gives a first, and satisfactory, interpretation of numbers raised to powers which are fractional, e.g. $2^{3\frac{1}{2}}$. From this interpretation a realistic understanding of logarithms could be developed.) (See *Number Appreciation,* chapter 5; from *Counting to Calculating,* chapter 10.)

The pattern of the above equations, if extended further, would give negative indices:

| on the left side (indices decreasing by one's) | on the right side (dividing by 10 each time) |
|---|---|
| $10^{-1}$ | $\frac{1}{10}$ |
| $10^{-2}$ | $\frac{1}{100}$ or $\frac{1}{10^2}$ |
| $10^{-3}$ | $\frac{1}{1000}$ or $\frac{1}{10^3}$ |

from which the induction is made that

$$10^{-4} = \frac{1}{10^4}$$

and so on.

If we incorporate the negative indices in the graph shown in fig. 27, the result looks like fig. 28.

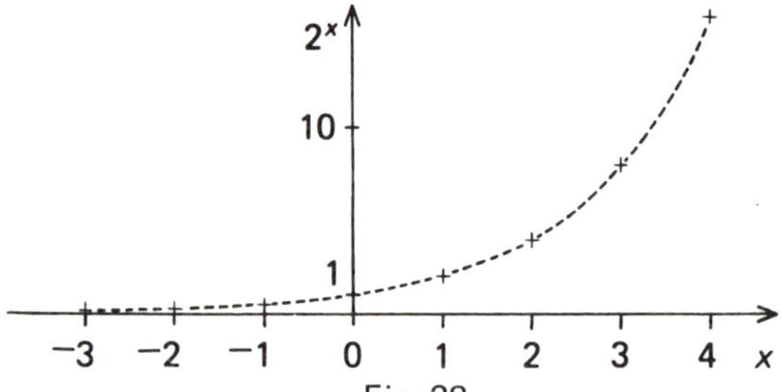


Fig. 28

The graph appears to continue to the left so as to approach the horizontal axis for negative indices. This supports the surmise suggested by the pattern.

## Brackets

Many formulae contain brackets, but the presence of brackets is intended to help and not to hinder, and when we need to substitute numerical values for symbols we normally have little difficulty in doing so. For example, we may be asked to find the Centigrade temperature corresponding to 68 °F, the formula being

$$C = \frac{5}{9}(F - 32).$$

Substituting 68 for $F$ we have

$$C = \frac{5}{9}(68 - 32)$$

$$= \frac{5}{9} \times 36$$

$$= 20.$$

But there are problems (often, it is true, of an abstract kind which would certainly not be suitable for many of our pupils) where computation is not involved, but where brackets containing letters require manipulation. In such cases it may seem simple to follow instructions and rules, but when the reasons behind these are not clear the blind attempt to follow them is accompanied by a precarious trust in memory rather than by an understanding of meaning, and it is no wonder that mistakes are frequently made.

There are several types of problem involving the 'removal of bracket's, namely:

(1) $a \pm (b \pm c)$

(2) $a(b \pm c)$

(3) $(a \pm b)(c \pm d)$

together with combinations and elaborations of these.

One very common misconception which pupils acquire and cling to is that *because* the combinations of signs used in the expansion of the above kinds of expression produce the same results as when the rules for directed number are applied, *therefore* the latter rules are the justification for using them.

For example, in simplifying or removing brackets from $a - (b - c)$ it is often assumed that $(-c)$ is multiplied by a factor $(-1)$ implied in front of the bracket; or more generally that in, say, $3(4a + b) - 2(a - 3b)$ the final term is $(+6b)$ *because* $(-3b)$ and $(-2)$ are directed quantities; and that a similar reason explains why, say, $(p - 3)(p - 5)$ yields a term $(+15)$.

But teachers at least should be aware that it is not *necessary* for directed quantities to occur in such expressions and that, therefore, there must be other ways of justifying the manipulations. We show how this might be done on pp. 64–68.

We now consider how the derivation of the rules for some of these types might be presented.

(1) $a \pm (b \pm c)$

It is possible to devise situations such as money transactions which can be expressed symbolically in different ways to show the identity of two different forms of expression. The illustration outlined below is mainly for the consideration of teachers, and is not advocated as a first presentation to pupils. It might be helpful at a later stage as 'corroborative evidence'.

Take, for example,

$$a - (b - c).$$

Suppose John goes out with 15 pence in his pocket; he buys a toy for 8 pence and offers a tenpenny piece for it, receiving 2 pence in change.

His spending can be expressed by the bracket $(10 - 2)$, and he leaves the shop with 7 pence,

so $$7 = 15 - 8 = 15 - (10 - 2).$$

But alternatively he may first go out with 15 pence, spend 10 pence in a shop — so that he has $(15 - 10)$ pence left — and then meet a friend who gives him 2 pence, so that again he finishes up with 7 pence, i.e. $15 - 10 + 2$ pence.

Both expressions produce the same outcome, and so

$$15 - (10 - 2) = 15 - 10 + 2.$$

This does not *prove* anything, but makes it 'look as if' the negative sign in front of a bracket is an instruction to change the negative sign inside to a positive when the brackets are taken away.

This way of looking at the negative sign as giving an instruction is a helpful as well as a permissible interpretation of its function since we are aware that the sign does have different meanings in different circumstances. When, for instance, we describe a temperature of 7 degrees below zero by the directed number $(^{-}7)$, the sign does not mean 'subtract'. (See chapter 4 of *Number Appreciation.*)

It must be admitted that it is hard to make demonstrations such as the above really convincing and they are more likely to satisfy the teacher or the mature student rather than one who is meeting the operation for the first time. Hence the reservation made earlier.

A more satisfactory way of dealing with this, the most difficult of these sign combinations (the others can be treated in a similar fashion) is suggested in chapter 7 of Sawyer's *Vision in Elementary Mathematics* (Pelican), and this is much more acceptable and satisfying to pupils. The method illustrates again the value of inductive reasoning, and goes as follows:

Consider the expression $25 - (20 - n)$.

The following table can be built up by asking members of a class or group to suggest various values for $n$ and then to evaluate the succeeding lines:

| $n$ | 2 | 5 | 7 | 8 | 10 | 13 | 15 | 18 | 19 |
|---|---|---|---|---|---|---|---|---|---|
| $20 - n$ | 18 | 15 | 13 | 12 | 10 | 7 | 5 | 2 | 1 |
| $25 - (20 - n)$ | 7 | 10 | 12 | 13 | 15 | 18 | 20 | 23 | 24 |

The first and third lines are now examined (on a blackboard the middle line could be rubbed out):

| $n$ | 2 | 5 | 7 | 8 | 10 | 13 | 15 | 18 | 19 |
|---|---|---|---|---|---|---|---|---|---|
| ? | 7 | 10 | 12 | 13 | 15 | 18 | 20 | 23 | 24 |

Pupils will soon see a relationship pattern between the lines; the lower line numbers are each 5 more than those immediately above them, so they can be described as $n + 5$ or $5 + n$.

Hence $25 - (20 - n)$ is the same as $5 + n$, and since 5 is equal to $25 - 20$, we can write

$$25 - (20 - n) = 25 - 20 + n.$$

Other expressions of a similar kind should be tabulated in the same way – individuals or groups could make up their own – and only then, when each equality has been listed and examined, would an induction be made which expresses verbally the rule to be followed.

Teachers will appreciate the corroboration afforded fig. 29, but it may be too sophisticated for many pupils.

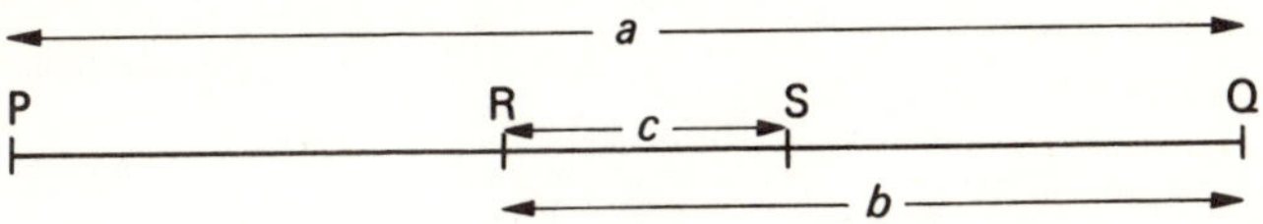


Fig. 29

Suppose PQ = $a$, QR = $b$, RS = $c$ units of length.

The length of PS can be expressed in two ways, viz. PQ − SQ and PR + RS.

Now SQ = $b - c$; PQ − SQ = $a - (b - c)$

and PR = $a - b$; PR + RS = $a - b + c$

and each of these expresses the length of PS, thus justifying

$$a - (b - c) = a - b + c.$$

(2) $a(b + c)$

Again, the mere *instruction* that this means 'multiply each term in the bracket by the quantity outside' is the reverse of a meaningful learning process by pupils, whereby the verbal description of a situation can be expressed in the alternative forms $a(b + c)$ and $ab + ac$.

Teachers will note that the relation is exemplified by the standard way of performing long multiplication.

Thus 37 x 23 is set out as:

```
  37
x 23
----
 740
 111
----
-------
----
```

so that it is equal to 20 thirty-sevens and 3 thirty-sevens

$$37 \times 23 \text{ or } 37(20 + 3) = 37 \times 20 + 37 \times 3.$$

In a more general way we can say that when we multiply a number $n$ by 23, we know that the result is $20n$ added to $3n$

$$n(20 + 3) = 20n + 3n.$$

Another step in generalization, where the multiplier can be written as $p + q$, then yields the statement

$$n(p + q) = np + nq$$

This is an expression of the familiar 'distributive law' discussed in greater detail in chapter 2 of *Number Appreciation.*

A more helpful introductory illustration for pupils, which gives the identity in a natural and inevitable way, is to consider the products as representing areas.

Suppose a school has two classrooms which can be thrown into one by removing a partition (fig. 30).

The combined area of the two rooms is

$$(9 \times 13) + (9 \times 11)$$

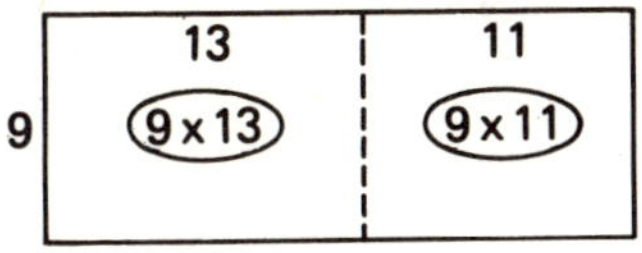


Fig. 30

whilst as one large room it would be written as

$$9 \times 24 \text{ or } 9 \times (13 + 11).$$

Again, the same steps can be taken to generalize this using letters, as

$$a(b + c) = ab + ac.$$

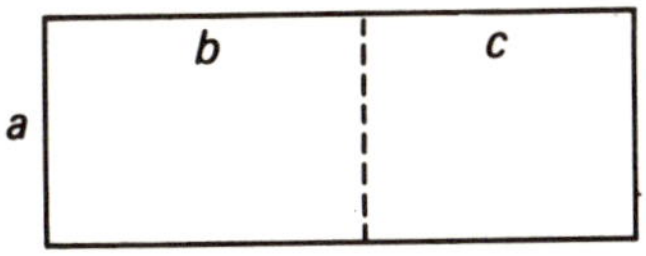


Fig. 31

Thus, a rule for removing brackets has emerged which practice with further examples will consolidate.

Similar 'two-way' representations can be obtained from shapes which gradually increase in complexity, provided they can be split into rectangles and reassembled in one long rectangular shape. A few examples are given which, as we mentioned earlier, should be each preceded by calculations using numbers, and then perhaps with the common width alone as a number before the final literal expressions are set up, e.g., for fig. 32(a) we might have successively:

$$A = (3 \times 4) + (3 \times 6) + (3 \times 5) = 3 \times (4 + 6 + 5)$$

$$A = (3 \times b) + (3 \times c) + (3 \times d) = 3 \times (b + c + d)$$

$$A = (a \times b) + (a \times c) + (a \times d) = a(b + c + d)$$

We may treat the shapes in (b), (c) and (d) of fig. 32 in a similar way.

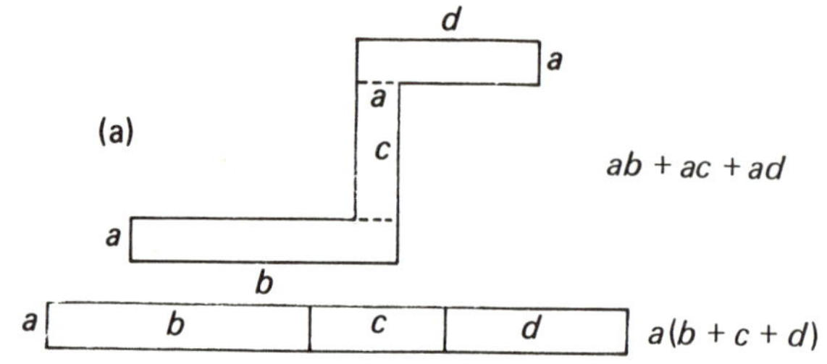


$\therefore a(b + c + d) = ab + ac + ad$

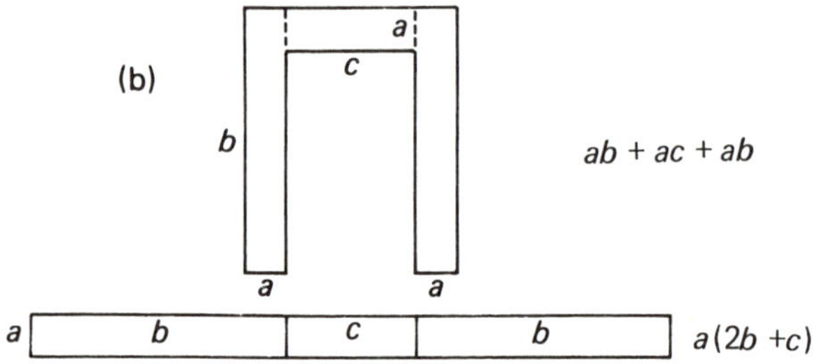


$\therefore a(2b + c) = 2ab + ac$

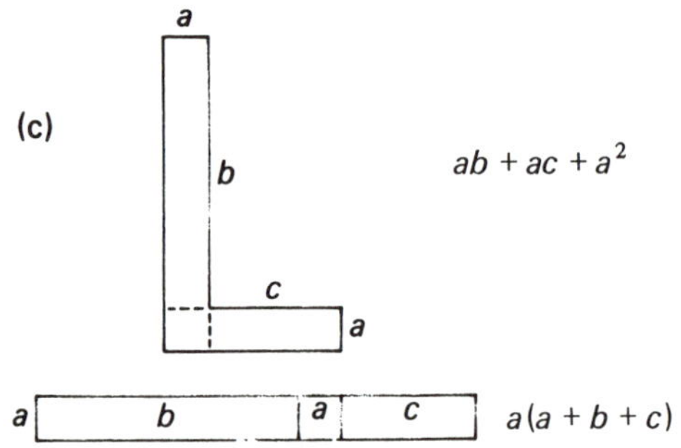


$\therefore a(a + b + c) = a^2 + ab + ac$

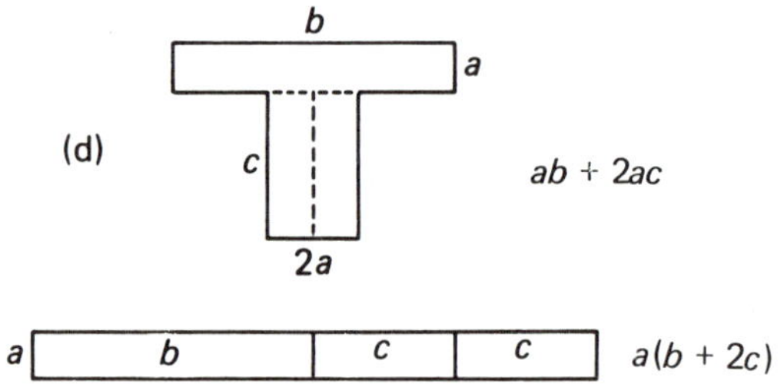


$\therefore a(b + 2c) = ab + 2ac$

Fig. 32

Clearly it would be helpful to precede such exercises as the above with equivalent activities involving pieces of structural equipment (Dienes, Tillich Blocks, etc.)

(3) $(a \pm b)(c \pm d)$

The expansion of this type of expression – including special cases such as $(a \pm b)^2$ or $(a + b)(a - b)$ – is generally made entirely dependent on techniques of manipulation which, however logical they may be to the mature mind, are rules of thumb without meaning to those pupils who are merely given instructions to memorize.

Once again, alternative representations of the areas of rectangles afford one way of showing the identity of different forms of expression, and it is more important than before that the rectangles should be in material which can be handled and not just seen as drawings on paper or blackboard. Dienes Algebraic Experience Material (AEM) and similar apparatus are most valuable, but the following suggestions make use of the simplest possible material, namely rectangular sheets of paper.

(i) $(a + b)(c + d)$

A rectangular sheet of paper is folded parallel to two adjacent edges to make creases (shown as broken lines in fig. 33) at right angles to each other. The lengths and breadths of the rectangles so formed are labelled as in fig. 33.

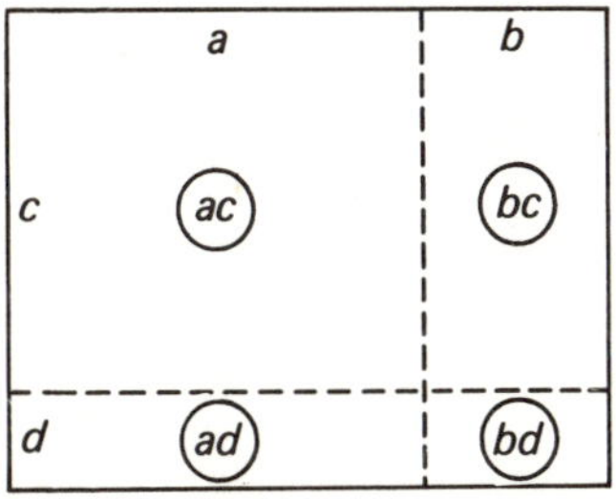


Fig. 33

The total area of the sheet is represented by the expression $(a + b)(c + d)$, and pupils will write in each of the four smaller rectangles the appropriate area.

It therefore follows that

$$(a + b)(c + d) = ac + bc + ad + bd.$$

Note how we may use this method for performing long multiplication, working on graph paper on which unit squares are divided into tenths.

For example,

$36 \times 28 = (30 \times 20) + (30 \times 8) + (6 \times 20) + (6 \times 8)$. (fig. 34.)

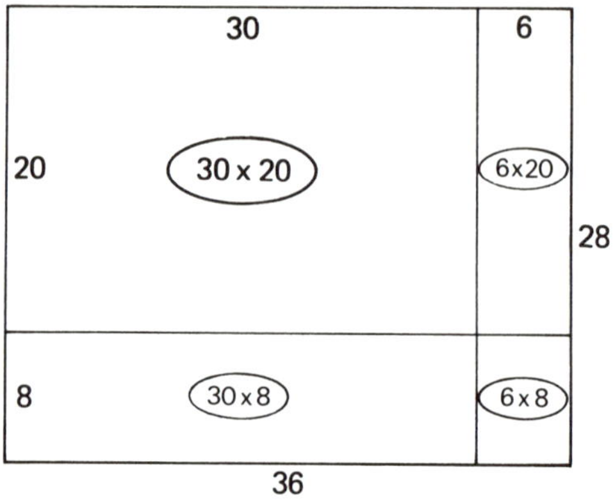


Fig. 34

The expansion of $(a + b)(c + d)$ gives a very general result. It can be followed up with other examples of which the following are typical:

$(a + 3)(a + 5)$

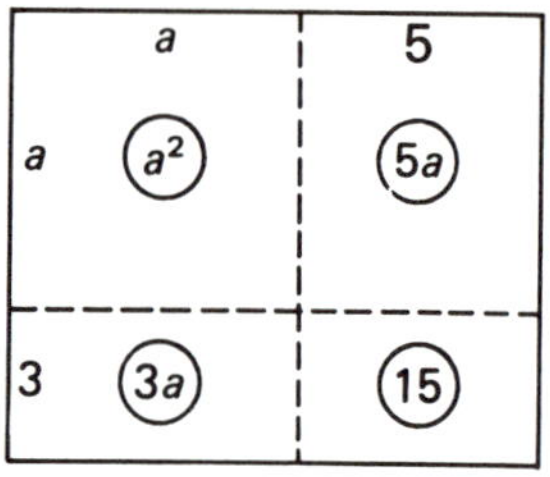


$$a^2 + 5a + 3a + 15$$
$$= a^2 + 8a + 15$$

Fig. 35

$(a + b^2)$

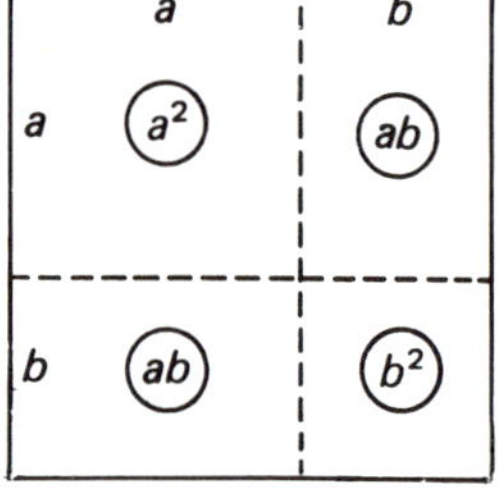


$$a^2 + ab + ab + b^2$$
$$= a^2 + 2ab + b^2$$

Fig. 36

$(2a + 3b)(a + 2b)$

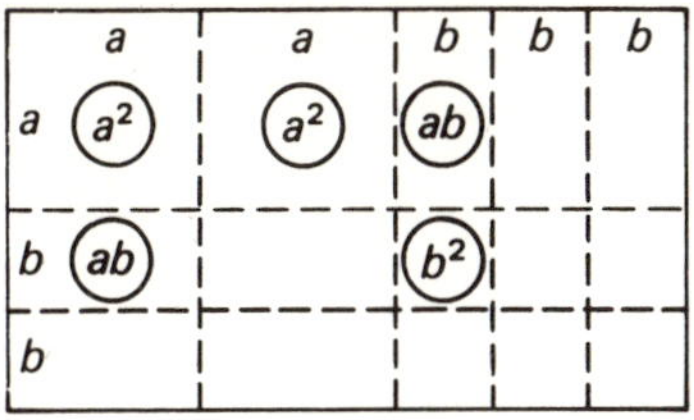


Fig. 37

$$a^2 + a^2 + ab + ab + \ldots + b^2 + \ldots$$
$$= 2a^2 + 7ab + 6b^2$$

When the method has been well practised, all the identity statements should be collated and examined, so that it can be ascertained how the terms have arisen. The induction will be made that 'it looks as if' each term in one bracket is paired in multiplication with each term in the other bracket.

Thus, just as in number work children discover, say, the equivalence of particular pairs of fractions *practically* through apparatus and drawings and later use their results to make a *mental* discovery of the way these pairs are related, so here the results of bracket manipulation first performed practically are used to make an induction of a mental procedure which replaces the diagrammatic method.

This is a general principle of learning which is used extensively in primary education and should find continued application amongst secondary school pupils.

(ii) $(a + b)(c - d)$

(The details of the procedure described below are much less clear if they are followed by the reader merely on the diagrams and a sheet of paper should be folded and cut as the text is read. Pupils must also use sheets of paper.)

A rectangular sheet is used, the pupil making two creases parallel to adjacent edges. This time, however, the whole of one side represents the length $c$, and only part of it is $d$ (fig. 38(a)).

(a)

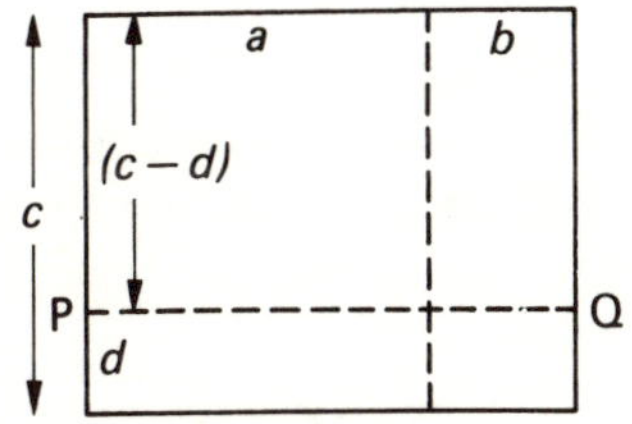


Fig. 38

The paper should be folded about PQ with the lower strip turned under, the top face then showing as in Fig. 38(b). This should then be shaded so that the shaded area represents $(a+b)(c-d)$.

After unfolding, the whole of the original sheet is now looked at to see how this shaded area is made up. Take the left-hand part, shown in

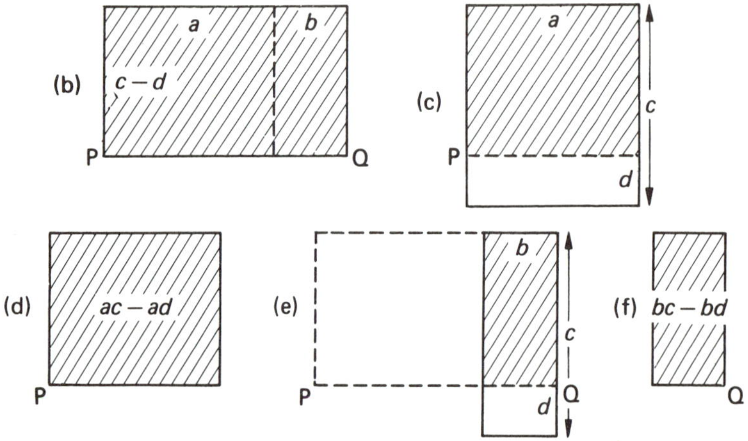


Fig. 38(b)–(f)

Fig. 38(c) (the other part should be folded so that it is out of sight). Its area is represented by $ac$.

Tear off the lower left-hand rectangle, whose area is $ad$, so that the remaining portion, which is part of the shaded area we need, has an area $ac-ad$, illustrated in Fig. 38(d).

Now unfold the right-hand part, whose area is $bc$ (fig. 38(e)).

Cut or tear off the lower unshaded portion, area $bd$, so that the part left has an area $bc-bd$ (fig. 38(f)).

Open out the two parts to show the complete shaded section, which therefore has an area which can be expressed either as $(a+b)(c-d)$ or as $ac-ad+bc-bd$ (fig. 39).

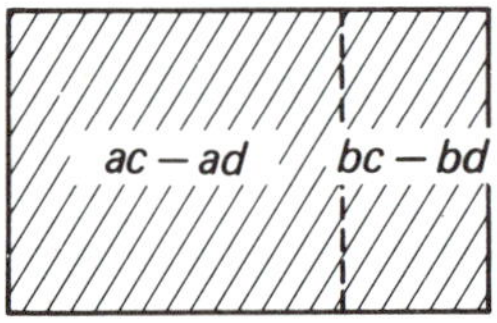


Fig. 39

(iii) $(a-b)(c-d)$

The expansion of this expression is likely to cause more difficulty than the expansion of any of the other types we are considering.

We shall suggest two slightly different ways of doing this here.

(a) The expression $(a-b)(c-d)$ may represent the area of a rectangle formed by cutting strips which are respectively $b$ and $d$ units wide, from a rectangle whose sides are respectively $a$ and $c$ units long (fig. 40).

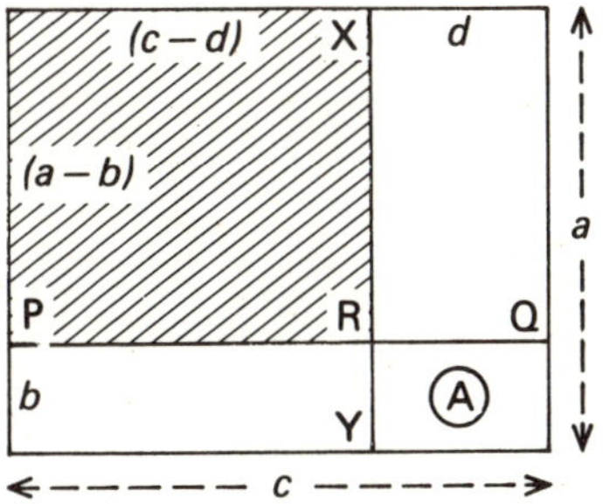


Fig. 40

The area of the shaded rectangle, which is clearly $(a-b)(c-d)$, can be found by beginning with the area $ac$ of the *complete* rectangle, and subtracting the *complete* strip across the bottom (area $bc$) and the complete strip down the right hand side (area $ad$).

How many pupils will spot the error made if we stop the operation here? Have we not *twice* subtracted the area marked A?

Having spotted our error (not our mistake) we can see that we can correct it by adding the area A to our statement. Thus, using symbols, we could write:

$$(a-b)(c-d) = ac - ad - bc + \text{area A}.$$

This method does not bring in the need for any knowledge of directed number operations. All that might be needed are preliminary build-up questions for pupils, such as:

Give a statement for the length of the line PQ ($c$)
Give a statement for the length of the line PR ($c-d$)
Give a statement for the length of the line XY ($a$)
Give a statement for the length of the line RY ($a-b$)

followed by similar questions on various areas in the figure.

It is important and instructive to remember that Greek mathematicians, who had no concept of negative quantities, were able to express such relationships as we are considering, even though they had no concise symbolism for them.

(b) It may be that some pupils will find it difficult to visualize the double subtraction and compensatory addition of demonstration (a). In this case we may vary the demonstration. We take a rectangular sheet of paper and use it to justify the building up of each stage in the symbolic representation. At each step pupils must calculate the area of the paper which is visible and must express it in symbols. The steps, which are in a different order from before, are:

(1) Show the whole sheet, which has an area $ac$ (fig. 41(a)).

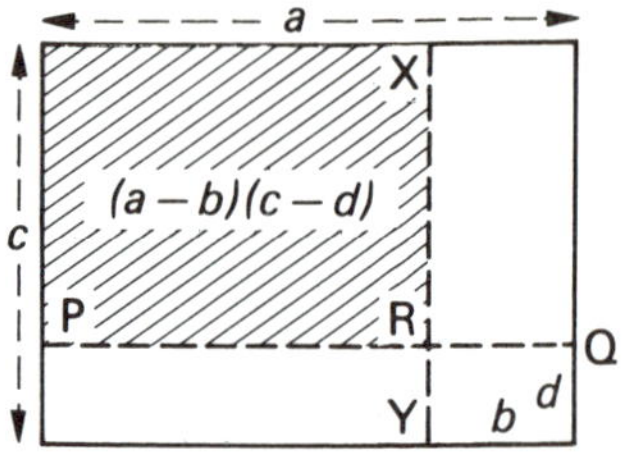


(a)

(2) Fold under the lower rectangle $ad$ so that it is out of sight. The area of paper now visible is $ac - ad$ (fig. 41(b)).

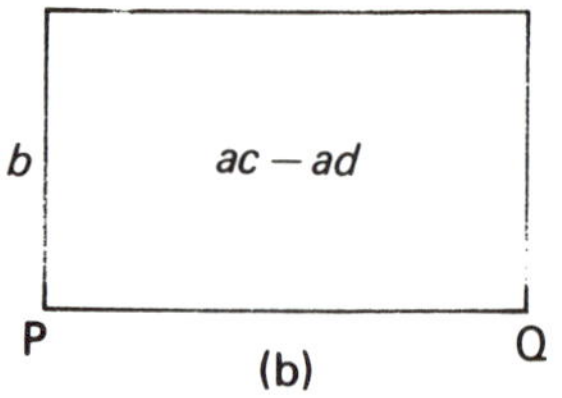


(b)

(3) Tear (or cut) along the line RY and bring back to view only the right-hand lower rectangle $bd$ so that the area now visible is $ac - ad + bd$ (fig. 41(c)).

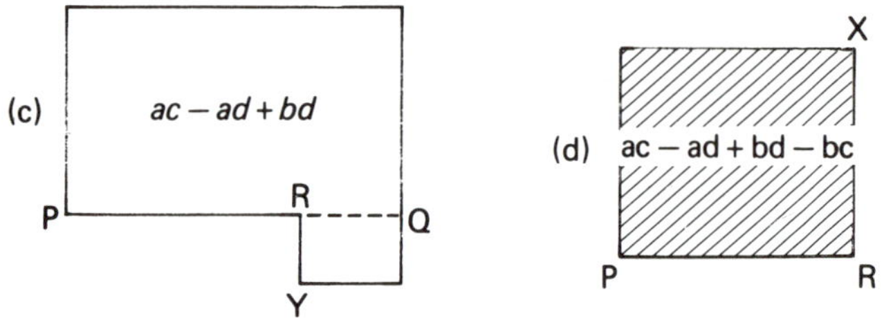


Fig. 41

(4) Cut off the right-hand strip, area $bc$, leaving a rectangle whose area is therefore $ac - ad + bd - bc$ (fig. 41(d)).
This rectangle, however, has sides $(a - b)$ and $(c - d)$ (fig. 41(a)).

Hence $\qquad (a - b)(c - d) = ac - ad + bd - bc.$

Because each stage in the procedure is made visible in a way which is not possible in a sketch, and because the areas successively shown are fully described by symbols, this method is more likely to be convincing and win acceptance. It will be noted that the crucial step is number (3) where the *addition* of the area $bd$ actually takes place, visibly and tangibly.

Some pupils find a 'guess and try' numerical approach to this product both interesting and convincing. The suggestions made below may appeal to them more than the demonstration by areas, whilst for others it might furnish another piece of supporting evidence.

The product of 28 and 17 would be found by most pupils by the standard method for long multiplication, and is 476.

It could also be written as

$$(30 - 2)(20 - 3)$$

If pupils have understood from multiplying out $(a + b)(c + d)$, or from multiplying our numerical examples such as $(30 + 8) \times (20 + 6)$, that pairs of terms have to be multiplied together, it is likely that they will do this first, obtaining $30 \times (20 - 3) = 600 - 90$, and $(2 \times 20) = 40$, which must also be subtracted.

The products $600 - 90 - 40$ are fairly 'obvious' guesses to write down, but what sign should be attached to the product $2 \times 3$?

If the statement is written

$$28 \times 17 = 476 = 600 - 90 - 40 \ ?\ 6$$

the next lines read

$$476 = 510 - 40 \ ?\ 6$$

and

$$476 = 470 \ ?\ 6$$

and the positive sign is clearly necessary in place of the question mark.

Several other products worked out in a similar fashion give an adequate basis for induction.

When sufficient experience has been gained in the above three kinds of multiplication (where numbers may be substituted for some of the letters, e.g. $(a + 4)(a - 3)$; or more general expressions expanded, e.g. $(2x \pm y)(x \pm 3y)$, all the identities should be collected, as suggested earlier. The operations and rules which are in future to be applied mentally are then induced from the identities themselves, the conclusion being that 'it seems to work *this* way'; it is then quite legitimate for the procedure to be crystallized as a technique. What is not educationally justifiable is for it to be given at the outset just as a series of instructions. We are not claiming that pupils will always remember the justification of a process. They will – and should – have recourse to rules but it is better that they should have seen their reasonableness even if they do not comprehend fully, than that they should merely be told what to do.

(iv) $(a-b)^2$

When $c$ and $d$ are respectively equal to $a$ and $b$ the special identity $(a-b)(c-d) = ac - bc - ad + bd$ becomes:

$$(a-b)^2 = a^2 - 2ab + b^2.$$

(v) $(a+b)(a-b)$

This is another special case which may be studied at this stage in the investigation or which may be deferred if factorization is going to be dealt with. Since it is more common for it to occur and to be applied as an alternative expression for 'the difference of two squares', we shall consider it in a later section (see p. 70).

### Factors

*Considerable doubts have already been expressed about the wisdom of attempting to follow the conventional 'headings' of an algebra syllabus with our pupils. It must be emphasized again that the content of this chapter is not put forward as necessarily desirable, and material is included which it would probably be wrong to advocate, even for pupils who are a little above average in ability. Sometimes, however, external demands are made with which teachers have to comply, and it has therefore been felt right to discuss content and its presentation which is really out of place for most of the pupils of the range of ability we are dealing with.*

This reservation applies very strongly to this section. Finding the factors of algebraic expressions has practically no utility value for our pupils, and it is hard to find any justification for their learning the processes. For a very few pupils there are certain features with which, it may be thought, they should have some acquaintance, and these will now (somewhat reluctantly) be considered. They are:

(1) Common factors.
(2) Difference of two squares.
(3) Complete (or perfect) square trinomials (with the allied process 'completing the square', as this may be needed for the solution of quadratic equations).

No justification can be offered here for the inclusion of any other features. The conventional trinomial, such as $x^2 - 6x + 8$ or $2a^2 - 5a - 3$ is commonly thought of as being important because of its application in the solution of equations. Since, however, only a sub-set of equations can be solved by factorizing and recourse otherwise has to be made to the standard formula, or to 'completing the square' (from which the standard formula is derived) it is difficult to find sound mathematical or educational reasons for making the majority of pupils struggle so uncertainly with this type of expression.

Teachers will realize that multiplication of bracketed expressions and factorizing are just different ways of looking at the same relationship.

For instance

$$a(b + c) \rightarrow ab + ac$$

is an 'expansion', while

$$ab + ac \rightarrow a(b + c)$$

expresses the sum of two terms as a product by finding the common factor.

With able pupils, in fact, the teacher would not separate the two techniques, but would try to ensure that they were understood and practised as alternative means of expression — just as, in fact, the primary school teacher stresses analogous relationships in number operations. For example, he will try to show that 7, 9, 63 are related by

$$7 \times 9 = 63; \;\; 9 \times 7 = 63; \;\; 63 \div 7 = 9; \;\; 63 \div 9 = 7.$$

With less able children, as already said, it is probably not wise to concentrate on the formal and abstract aspects of factorization, so the building up of such links will be harder to achieve.

It may be observed that one of the difficulties which pupils find in factorizing springs from a lack of real understanding of the words 'factor' and 'product'. These are terms which should have been in common and frequent use in numerical relations (see the section on Indices, p. 51). Unless the meaning of these terms is clear in those contexts, and unless pupils have become familiar with them, it is of little avail to talk about them in the field of algebraic expressions.

### 1. Common factor

From what has just been said, it is clear that the illustrations previously used for discovering a meaning for multiplying the terms in a bracket by a quantity outside are equally effective for the appreciation of the factorizing process.

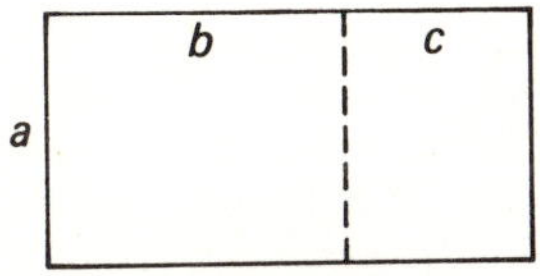


Fig. 42

To see by an area diagram (fig. 42) that $a(b + c)$ is the same as $ab + ac$ is equally to see that $ab + ac$ can be written as $a(b + c)$. Or, prefacing the generalization with a numerical example, that seven threes plus four threes is the same as eleven threes, i.e. (seven plus four) threes, and so on.

All the examples which were worked out for earlier practice on the expansion of brackets of this kind can therefore be re-examined from the opposite point of view and the mental process should then be generalized.

## 2. Difference of two squares

Examples 3 and 4 in chapter 2, which revealed interesting relationship patterns, could be looked at again and generalizations might be made to yield the standard identity

$$a^2 - b^2 = (a + b)(a - b).$$

A more direct way of seeing the relationship is as follows, using numerical examples to pave the way for an induction of the general statement:

Each pupil has a square card from which a smaller square has been cut. (It is advisable to have whole number dimensions.)

If for example the two squares have sides 9 units and 4 units, the area of the shape is 81 − 16, or 65, square units.

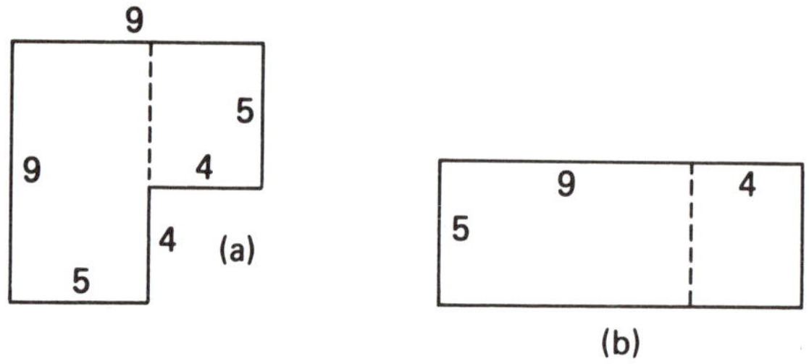


Fig. 43

The dimensions should be written in as shown in fig. 43(a).

The shape is then cut along the broken line into two rectangles which observation – and reasoning – will show to have sides of common length, 5 units, and so can be fitted together to make the one larger rectangle of fig. 43(b).

The area of this shape is the same as before, and it can be expressed as 13 x 5 square units

which is $$(9 + 4) \times (9 - 4)$$

and so $$9^2 - 4^2 = (9 + 4)(9 - 4).$$

It may be found that some pupils are still uncertain about the principle of 'conservation of area – this is not at all uncommon among lower ability pupils in the secondary school. If preliminary discussion shows that they have any doubts about this, then the pupils are not ready for the reasoning involved in the method.

The verbal relation can be translated into symbols after a number of examples have been recorded, and of course numerical practice, as in calculations like $101^2 - 99^2$, is necessary to show how the relationship can be put to good use.

The visual device described below is not at all suitable for an initial demonstration because it is rather too sophisticated, but it may be of interest later when the relationship is really understood.

A board (wood, plywood, etc.) of the shape representing $a^2 - b^2$ is cut into two identical pieces L and M along the broken line as shown in fig. 44(a). By turning M over and repositioning it, we form the shape of fig. 44(b). The same transformation could be achieved by having a hinge or pin suitably placed at P.

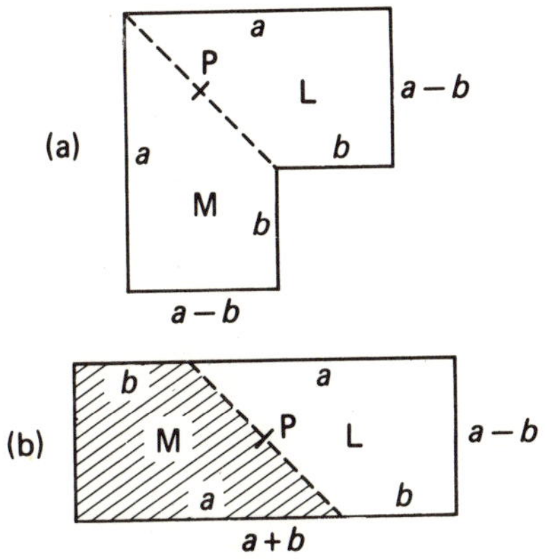


Fig. 44

In the first position the area of the board is seen as $a^2 - b^2$. In the rectangular position the sides are $a + b$ and $a - b$ and so the area of the rectangle is $(a + b)(a - b)$. We see that $a^2 - b^2 = (a + b)(a - b)$.

**3. Perfect square trinomials**

These, of course, are the results of expanding expressions of the type $(a \pm b)^2$, and the relationship should be recognizable in simple expressions in either form, although plenty of practice will have been needed before, say, $a^2 - 12a + 36$ and similar trinomials are so identified.

'Completing the square' is a process used in solving quadratic equations, and the standard rule for doing this may be only properly appreciated after it has been induced from area illustrations.

In its general form the question 'What must be added to $a^2 + 2ab$ in order to make it a perfect (or complete) square?' is seen to be answered by fig. 45. In practice, we can use cards or blocks of appropriate dimensions. (Teachers who use Dienes A.E.M. apparatus know how valuable it is for this purpose, and how adept pupils become at using it.)

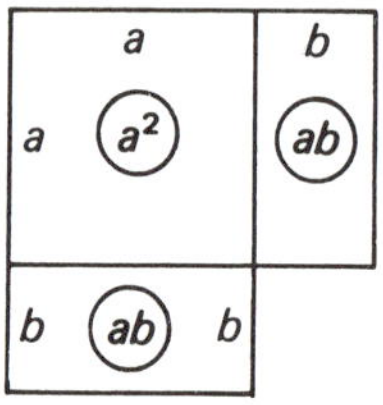


Fig. 45

If cards for particular trinomials are made, such as $a^2 + 8a$ with rectangles $a \times a$, $a \times 4$, $4 \times a$, and careful grading is organized, pupils will make the induction – although it will take time to reduce it to such a concise form as this – that *the square of half the coefficient of the term in 'a' must be added to complete the square.* It will be clear from the arrangement of the pieces of card what is the quantity (e.g. $a + 4$) whose square we have completed.

After this pupils will be ready to work abstractly, following formal rules, and using diagrams only as an occasional check. Again *practical* activity is followed by a *mental* process, but the latter has meaning only because the practical stage has been worked through.

The building up of a model or of a diagram when the term of the first degree is negative, as in $a^2 - 6a$, is by no means so simple, as will be appreciated from the method which has to be used to demonstrate practically the products $(a - b)(c - d)$ and $(a - b)^2$. It is legitimate however to use the clue given by the previous type-expression, where all the terms are positive, and to test in a purely manipulative way whether the same rule applies; and of course this is seen to be the case.

## Simple, linear, or first degree equations

In chapter 1 we gave illustrations of the way in which equations arose out of a certain use or interpretation of formulae. One example quoted was the application of the formula

$$F = \frac{9}{5}C + 32$$

to calculate the Centigrade temperature when the Fahrenheit reading was 68 °F.

On substituting 68 for $F$ we obtain the first degree equation

$$\frac{9}{5}C + 32 = 68$$

and we pointed out on p. 14 that this is no different in principle from an example such as

$$4x + 19 = 75.$$

Although the use of the formula may therefore involve the need to solve the equations, we do not suggest that this should necessarily be the initial motivation for investigating this particular type. One reason is that few formulae are so simple and straightforward that they can be used satisfactorily to introduce the topic; another is that if they are simple then verbal reasoning usually gives the required answer at least as readily.

For example, the formula $A = \frac{1}{2}bh$ need not be used to yield an equation to find the height of a triangle whose base is 8 units and area 40 square units. From the fact that the area of a triangle is half that of the rectangle it 'belongs to', one would argue that in this case the rectangular area is 80 square units; and since the base is 8 units, the height must be 10 units.

If formulae are not used, the difficulty then arises that it is not easy to find genuine and relevant problems which require formal treatment in an introduction to the study of equations, and recourse has to be made to artificial and abstract ones.

An essential preliminary – and it is a stage which is frequently overlooked or dealt with too superficially – is the translation of simple symbolic expressions into verbal statements and vice versa, so that the basic operations which are usually quite easily followed in words are recognized and appreciated in symbols. For example:

(1) $n + 5 = 9$

means 'when 5 is added to a number, the result is 9'. What is the number, and how is it obtained?

The verbal explanation is 'take 5 from 9' and symbolically this is expressed as

$$n = 9 - 5.$$

We have, therefore, two equivalent statements.

Similar translations should be made for, say,

(2) $n - 3 = 7$ e.g. Take 3 from a number and the answer is 7; so the number is 7 + 3)

(3) $4n = 12$ (4 times a number is 12; so . . .)

(4) $\frac{n}{5} = 2$ (one-fifth of a number is 2; . . .)

Teachers will realize that these are fundamental types encountered on the way to the solution of more general equations, and *verbal* understanding is a necessary factor in our pupils' ability to cope with them in a way transcending the merely mechanical.

It is also helpful to see how the examples given above are related to the basic number bonds, which themselves can be expressed in different ways. Thus:

$$9-5=4 \quad : \quad 4+5=9$$
$$7+3=10 : 10-3=7$$
$$4 \times 3=12 : \quad 3=12 \div 4$$
$$10 \div 5=2 \quad : \quad 10=2 \times 5$$

are some of the equations which can be related to the above four types.

When it is appropriate to deal with the more general type of linear equation, many teachers profitably use the 'balance' method to illustrate and emphasize the essential characteristic of an equation namely that whatever operation is performed on one side must also be performed on the other in order to preserve equality.

A pictorial method (or, indeed, a practical one using a laboratory balance) might show 3 bags, each containing the same number of nails, which together with 9 more nails just balance half a gross of nails on the other pan (with, perhaps three similar empty bags as well).

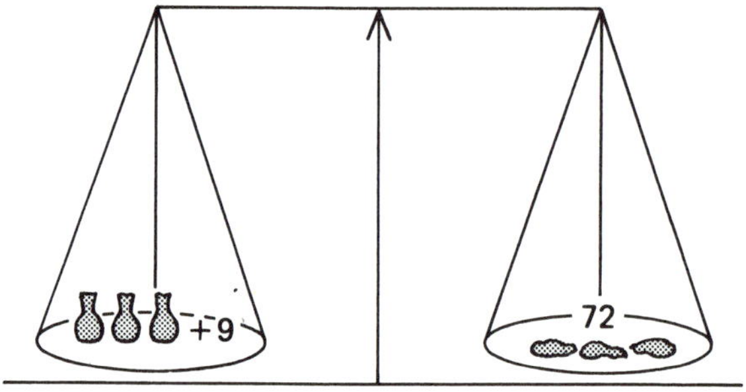


Fig. 46

If $n$ represents the unknown number of nails in each bag, the following equation can be made up:

$$3n+9=72.$$

The physical action of removing 9 nails from each side in order to preserve the balance (i.e. the equality) is expressed symbolically as

$$3n+9-9=72-9$$

whence

$$3n=63$$

and the term containing the unknown has been isolated.

As suggested above, it is now advisable to translate the next step into words: 'Three times a number is 63; what is the number?' rather than pose a question such as 'What must we do to $3n$ to get (one) $n$?' which often seems to confuse pupils unduly.

A number of such problems and equations treated similarly provide the basis for a logical procedure.

It is necessary, however, to proceed with caution when an equation of the type $5x - 8 = 72$ is introduced. Here the number 8 has to be *added* to both sides, and very often pupils have difficulty in understanding this apparently simple step. This is perhaps because it is not so easy to make up a convincing 'story' to illustrate the situation. It may also be because teachers often fail to appreciate that there *is* difficulty since they are unconsciously using their own knowledge of directed number in seeing that the addition of 8 to the left-hand side $(5x - 8 + 8)$ isolates the $5x$ term as required.

The illustration of the bags of nails can perhaps be adapted as follows:

'There are 4 full bags and one from which 8 nails have been taken; these balance half a gross in the other pan.'

(One must take care not to leave a negative number of nails in a bag!)

Then the equation becomes

$$4x + x - 8 = 72$$

or

$$5x - 8 = 72$$

If the 8 nails are replaced this is equivalent to adding 8 to each side of the balance/equation, and since there are now 5 full bags this gives

$$5x = 80$$

which is the same as $5x - 8 + 8 = 72 + 8$.

In due course the operations of subtracting from, adding to, and dividing each side by the same quantity will be recognized as the standard procedures to be followed. The language used may be that of 'opposites'. If 8 is already added to the term to be isolated, then the 'opposite' process of subtraction must be performed on each side; if the unknown is *multiplied* by 5, then each side must be *divided* by 5 (thus: $5x \div 5 = 80 \div 5$); and so on.

This is, of course, akin to the modern concept of 'inverses' which many teachers have readily accepted and which they apply to the solution of equations, but which others feel unnecessarily diffident about using. However, as long as the idea of opposites is not applied blindly merely as a rule, but adequate illustrations are given and verbal translations are made both from and into symbolic expressions the terminology is legitimate and helpful.

With all equations, and by whatever method they are solved, it should be an accepted routine to check solutions to ensure that they are correct.

We now describe another approach to equation solving which has the merit of sheer interest in its initial presentation, so that the question of artificiality does not in practice arise. It also illustrates once again the value of inductive thinking, as contrasted with the deductive basis of the above method.

The teacher may say: I think of a number; I double it, and add 5; the answer is 17. What is my number?

Such an example, followed by similar ones – including some where a number is subtracted after an initial multiplication or division – bring pupils gradually to realize, and to be able to explain, that the mental steps required to solve the puzzle reverse (or 'undo') the operations which were originally described. And the children themselves are usually eager to make up puzzles for other members of the group to solve.

The above, which is all oral and mental, is the first stage.

Next, each of the verbal puzzles is translated, phrase by phrase, into shorthand symbols. For instance, 'I think of a number' can be replaced by '$n$'; 'I double it' gives '$2n$' (or '$2 \times n$'; the notation depends on the degree of experience pupils have had in symbolic representation. It is clear that number puzzle equations can be introduced informally quite early to children). 'I add 5' leads to '$2n + 5$'; 'the answer is 17' gives finally '$2n + 5 = 17$'.

A set of similar equations results from the translation of each puzzle posed by teacher and pupils.

The original puzzles are then examined once again, this time so that pupils can say what the first step was in mentally solving each of them and to express this solution in a shorthand way.

Thus, in the example given, we first subtract 5 from 17, and this we know then to equal twice the number.

Accordingly one translates and writes:

$$2n = 17 - 5$$

and similarly for all the others.

When all the puzzles have been discussed in this way, the teacher (and if so desired, the pupils) will have two lists of shorthand statements which should be brought together in parallel sets.

They may run as follows:

| | |
|---|---|
| $2n + 5 = 17$ | $2n = 17 - 5$ |
| $3n + 2 = 14$ | $3n = 14 - 2$ |
| $2n - 3 = 21$ | $2n = 21 + 3$ |
| etc. | etc. |

An examination of these pairs of statements, each member of a pair having been obtained independently, yields the induction that 'it looks as if' the transfer of a number from one side of an equation to the other always results in an 'added' number becoming a 'subtracted' one and vice versa. This hypothesis will be tested in the solution of new puzzles (i.e. equations) and will become an accepted working rule.

From this point onwards, well-graded examples involving perhaps answers in fractions, use of brackets, fractional coefficients, the unknown on both sides of the equation, and so on, will take the study of these equations as far as will be appropriate. In dealing with all such types there is considerable value in continuing to stress the equivalence of verbal and symbolic statements; frequent practice should be given not only in devising puzzles and problems which could be translated into symbolic forms, but also in making up the puzzle or situation which any particular equation could represent.

As always, there is no question of *either* one particular approach to the solution of equations *or* another. It is far better to help pupils see that more ways than one are possible, although time and practice must be given for real understanding. Quite often too it will be found that a pupil who has had difficulty with one approach will find a fresh line of attack more meaningful and will regain confidence.

Much might be said about formal methods of solving equations; it must be remembered that these methods are not important as ends in themselves. Teachers are well aware of the 'conditioning' to which they were probably once subjected, which made them react with techniques; this is justifiable when economy of time is desirable and when the techniques do not conflict with understanding. But if pupils can interpret equations (even in examination papers) as number puzzles and solve them mentally that way, clearer thinking is going on than when rules are mechanically applied and such thinking should be vigorously encouraged.

**Simultaneous equations**

The topic can be given concrete and pictorial meaning if it is approached through the study of the linear function and its representation by the straight-line graph. It seems rational to bring together in this chapter the different kinds of equation which will be studied by at least some of our pupils. (Only simultaneous equations of the first degree will be considered.) The approach suggested will pre-suppose a knowledge of the straight-line graph and its formula. Implicit in the development is the realization that a formula is a *relationship* between variables in which one changing quantity causes another, which depends on it, also to change (see *Some Simple Functions*).

The following example will show how the problem may be introduced:

Water runs into a tank A, initially containing 12 litres, at the rate of 3 litres per second, whilst water runs out of another tank B, initially holding 92 litres, at 5 litres per second. (The inlet and outlet pipes start to operate at the same instant.)

Graphs of the situation are drawn from the above data (a sketch only, not drawn to scale, is given here (fig. 47) but graphs on a squared board should be used for actual illustration and discussion). For A, the starting point is 12 on the quantity axis, and for each second there is a rise of 3 units; for B, starting at 92, the graph falls 5 units for each horizontal unit of time.

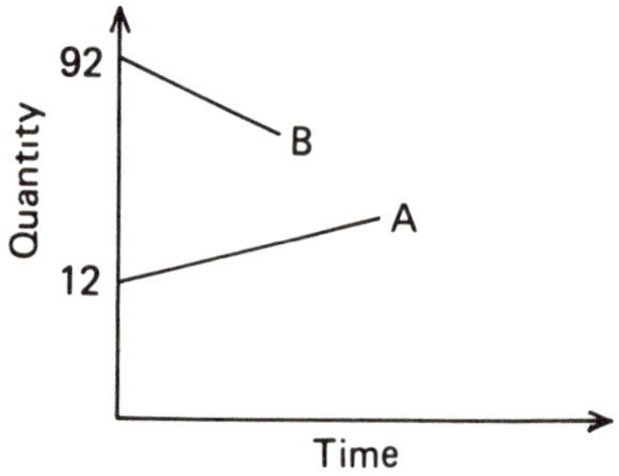


Fig. 47

Independently, the formulae linking quantity of water with time in each tank are built up; the amounts in tank A at the end of successive seconds are 12 + 3; 12 + 2 x 3; 12 + 3 x 3 etc. so after $t$ seconds the tank contains $12 + 3t$ litres. The argument for tank B goes similarly so the formulae are

$$\text{Tank A} \quad Q = 12 + 3t$$
$$\text{Tank B} \quad Q = 92 - 5t.$$

Pupils can be asked to agree that at an instant in time represented by point P in fig. 48(a) the broken lines PS and PT represent the quantities of water in the tanks. It is clear that at time P there would be more water in tank B than in tank A, and the readings on the graph show the respective quantities.

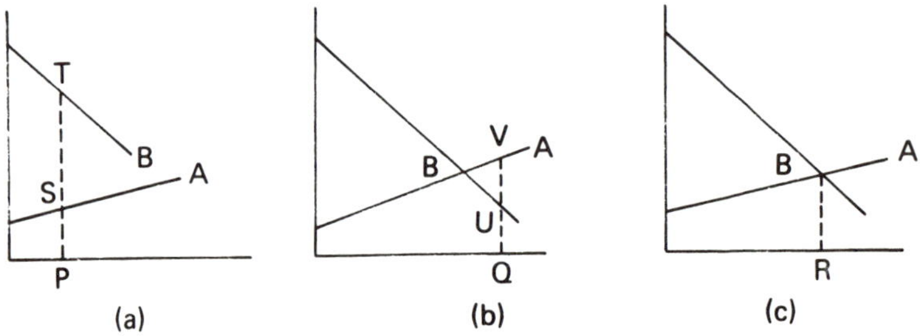


Fig. 48

Similar reasoning would apply to a point such as Q (fig. 48(b)), with QU and QV representing the quantity of water in each tank, and then R (fig. 48(c)) would be seen to represent the instant when both tanks contain the same amount of water.

The lack of precision in reading a graph would be a point for discussion, leading to the recognition of the further advantages of a 'mental' rather than a pictorial method for ascertaining the above information. It would have to be understood that at P and Q the length of a broken line is a value of either $(12 + 3t)$ or $(92 - 5t)$ and substitution of, say, 2 for $t$ at P and 15 for $t$ at Q would give the quantities by calculation instead of by reading the graphs.

If then the time at R were $n$ seconds from the start, the amounts of water in the two tanks would be $(12 + 3n)$ and $(92 - 5n)$ and these would be equal. We therefore have the equation

$$12 + 3n = 92 - 5n$$

which is of a familiar type.

A number of similar problems could follow, emphasis being laid on the correspondence between the lengths of lines in graphs and the values of the respective functions or dependent variables which are the subjects of the formulae. Once again, it is worth stressing that constant recourse to graphical sketches for illustration gives meaning to the calculation technique.

When in due course the more conventional type equations arise out of other kinds of problems, or occur just as mechanical exercises, the emphasis should still be on relationships between variables even though it may be less obvious that these are inherent in the situation.

The problem may be entirely an abstract one, as in the instruction 'Solve the simultaneous equations $5x + 2y = 50$; $3x + 4y = 44$.' (The example here chosen is a more general one than would be selected for pupils at an early stage in the progressive development. A note is appended below on the need for grading.) In interpreting such equations pupils should be helped to understand that the set of pairs of values of $x$ and $y$ which satisfy the first is infinite in number, e.g. $(1, 22\frac{1}{2})$; $(2, 20)$; $(4, 15)$; $(6, 10)$ etc. so that the relationship between $x$ and $y$ is a continuous one, and one which can be illustrated by a straight line graph.

The other relation is also represented by a linear graph and the process of 'solving' the equations consists in finding the pair of values which is common to them, i.e. the coordinates of the point of intersection of the two lines.

The graphs are best visualized when the relations are written in the forms,

$$y = -2\tfrac{1}{2}x + 25$$

$$y = -\tfrac{3}{4}x + 11$$

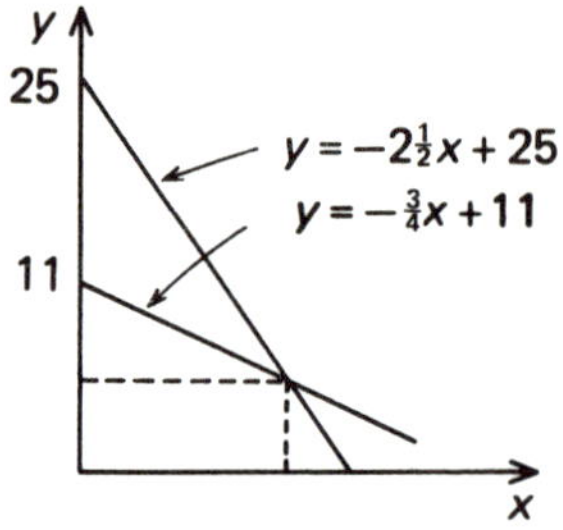


Fig. 49

the coefficients $-2\frac{1}{2}$ and $-\frac{3}{4}$ representing the respective gradients of each line.

In fig. 49 the solution is illustrated graphically, but by calculation it is a matter of finding when the '$y$' values are equal and this is done by writing

$$-\tfrac{3}{4}x + 11 = -2\tfrac{1}{2}x + 25$$

This gives $$1\tfrac{3}{4}x = 14$$

or $$7x = 56$$

$\therefore$ $$x = 8$$

and substitution yields $y = 5$.

These values would, of course, be checked in the original equations.

Or a problem may be stated in particular terms, such as the following:

In a youth-club coffee bar one group of members paid 34 pence for 5 cups of tea and 6 cokes. A neighbouring group paid 26 pence for 7 cups of tea and 3 cokes. What are the prices of a cup of tea and a bottle of coke?

If $x$ and $y$ are taken as the unknown prices in pence, the equations are just the same in principle as before, namely:

$$5x + 6y = 34$$
$$7x + 3y = 26$$

The graphs can be sketched roughly from the equivalent forms

$$y = -\tfrac{5}{6}x + 5\tfrac{2}{3}$$

$$y = -2\tfrac{1}{3}x + 8\tfrac{2}{3}$$

and the solution could be calculated by equating the values of $y$.

The amount of tedious arithmetic involved in the steps of such a solution is a stimulus towards the search for a simplification of the process, and discussion should eventually make it clear that if we multiply each term in an equation by the same number, this does not affect the equality of the two sides. Such a discussion should not rely on the theoretical argument, however obvious and reasonable it may be, for teachers are aware that by itself it is not inevitably fully convincing to pupils. Numerical checks make the conclusions of logic more satisfying to pupils, so it is well to illustrate this principle with copious examples until the principle is accepted. Thus the earlier equation

$$5x + 2y = 50$$

was known to be satisfied by

$$x = 6, y = 10.$$

If each term is doubled, giving

$$10x + 4y = 100$$

their substitution on the left-hand side yields 100 as it would be expected to do.

Hence, returning to the pair of equations

$$5x + 6y = 34 \qquad \ldots (1)$$
$$7x + 3y = 26 \qquad \ldots (2)$$

values of '$6y$' in each would avoid the intrusion of fractions, so the successive steps would be:

| | |
|---|---|
| From (2) | $14x + 6y = 52$ |
| $\therefore$ | $6y = 52 - 14x$ |
| From (1) | $6y = 34 - 5x$ |
| Hence | $34 - 5x = 52 - 14x$ |

and so on, making for a much simpler procedure.

It is much more worthwhile going through such detailed steps with the aim of helping pupils to discover, or at least to see the reasons for, the quicker method eventually adopted than to wrap up the technique from the start in a series of formal instructions.

It need hardly be said that for our pupils carefully graded examples would be required at this 'technique' stage to ease the way towards more general problems. For instance, there should be pairs of equations in which the coefficient of $y$, first in both equations and then in one of them only, would be unity. Later, one coefficient of $y$ would be chosen to be a multiple of the other coefficient (as above). Eventually the coefficients would be such that the terms in each equation would have to be operated on to make equal multiples of $y$, as in $2x + 5y = 16$. $5x + 3y = 21$. Grading would also have to cover directed numbers both as coefficients of the unknowns and as the number terms.

When in time we are able to dispense with graphical illustrations, the need to obtain equal $y$ coefficients disappears, and if it is more convenient to do so we can make the coefficients of $x$ equal instead. Thus $2x + 5y = 16$, $5x + 3y = 21$ can equally well be solved by changing them to:

$$10x + 25y = 80$$
$$10x + 6y = 42$$

in which case $80 - 25y = 42 - 6y$

which gives $19y = 38$.

The above is one method by which the steps leading to a traditional method of solving simultaneous equations could be rationalized, so that pupils see meaning and common-sense in the procedure. Of course there are other ways in which the topic could be developed; the purpose here has been to link it with the concept of relationship and the linear function.

**Quadratic equations**

*The limited field of application, at a simple level, of quadratic equations, the artificiality of many textbooks problems which lead to them, and the barren mechanics of the usual kind of exercises, should be sufficient reasons for making a teacher ponder hard before deciding whether their study can be justified for any but academically bright pupils following a prescribed syllabus.* Should it be deemed desirable or necessary to develop the topic, the study of formulae does at least provide a concrete basis for application and approach, and so long as pupils have had plenty of experience with graphs, and in particular with graphs of formulae, a realistic and common-sense development is possible.

On p. 13 we discussed briefly the formula $h = 200t - 4{\cdot}9t^2$ where $h$ is the height in metres $t$ s after an object has been shot vertically into the air at a speed of 200 m/s. This formula is a particular case of the general one

$$h = ut - \frac{1}{2}gt^2$$

where $u$ is the initial velocity of the object and $g$ is the acceleration due to gravity. When $u$ is measured in metres per second, $g$ is $9{\cdot}8$ m/s$^2$.

In order to simplify the calculation involved, we use here an initial velocity of 98 m/s, which gives the formula $h = 98t - 4{\cdot}9t^2$.

Using this formulae we can compile a table relating height and time:

| Time, $t$ (s) | 0 | 2 | 4 | 6 | 8 | 10 | 12 | 14 | 16 | 18 | 20 |
|---|---|---|---|---|---|---|---|---|---|---|---|
| Height, $98t - 4{\cdot}9t^2$ (m) | 0 | 176 | 314 | 412 | 470 | 490 | 470 | 412 | 314 | 176 | 0 |

We may draw a graph to represent the relationship (fig. 50). This graph is used for interpolation, i.e. readings for other values of the variables not given in the table.

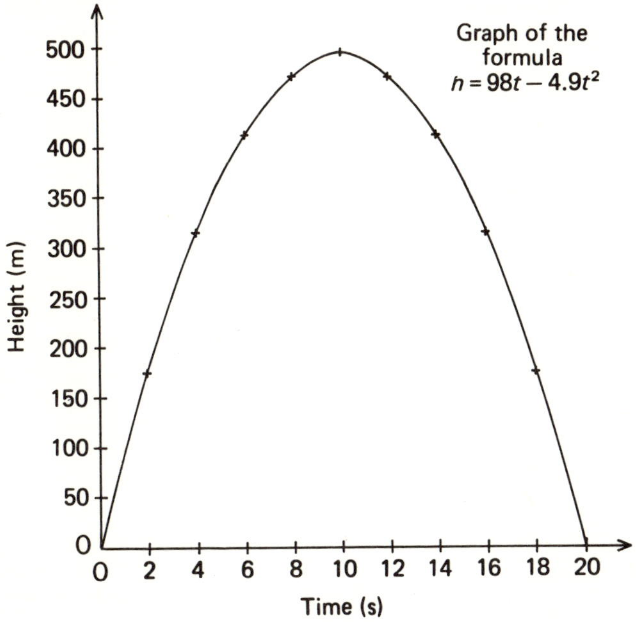


Fig. 50

From their experience of continuous graphs pupils should know that they can be used to find the answers to two kinds of questions. When they are graphs of formulae the answer to one kind is equivalent to making a straightforward substitution in the formula. The answer to the other kind is in fact the solution of an equation, and a graphical method will often be the best means available at a particular stage for any pupil, advanced or not. (It is not always appreciated that some equations cannot be solved by exact methods.)

In this case the questions are:

(1) What will be the height of the object for a stated value of $t$?
(2) At what time (or times) is the object a stated height above the ground?

First take questions of type (1). For example, how high is the object after (a) 3 s (b) 5 s?

The point of real importance here is that a group of pupils will almost certainly obtain from the graph a number of different estimates, and discussion should make them aware that probably no particular one is correct, but that all are approximate.

But if 3 (and then 5) is substituted for $t$ in the formula, only *one* result is obtained, and this will be accepted as the true one; the graphical estimates are different from one another because of the extreme difficulty of drawing and reading a graph with precision.

Now consider a question of type (2). How long elapses before the object is, say, 176 m above the ground? From their graphs slightly varying times will be obtained by pupils. (There are, of course, *two* answers to the question.)

The subsequent discussion must bring out and underline the fact that, as before, there ought to be a unique theoretical answer. This is the really crucial step which will motivate the development which is to follow.

Substituting 176 for $h$ in the formula gives

$$176 = 98t - 4{\cdot}9t^2$$

and this asks the question 'What value of $t$ makes $98t - 4{\cdot}9t^2$ equal to 176?' Here is a type of equation which has not been met with before; how can it be solved? *This is the new problem.*

The equation can be written

$$4{\cdot}9t^2 - 98t + 176 = 0$$

or

$$t^2 - 20t + 36 = 0$$

after dividing through by 4·9, and the new feature it contains is $t^2$, the square, or second degree term of the unknown quantity, $t$. (Note that here we are making an approximation to ease the subsequent burden of calculation: to one place of decimals 176 ÷ 4·9 gives 35·9.)

**A short digression**

If we had such an equation in which the first degree term, in this case $20t$, were absent the solution would be quite simple.

For example, the question 'What is the length of the side of a square plot whose area is 81 $m^2$?' can be written as the equation $s^2 = 81$ and it is solved by finding the square root of 81.

The problem can be taken a stage further as follows:

'What is the side of a square plot, if, after an L-shaped path 2 m wide has been cut from it, a square of area 81 $m^2$ is left?' (fig. 51)

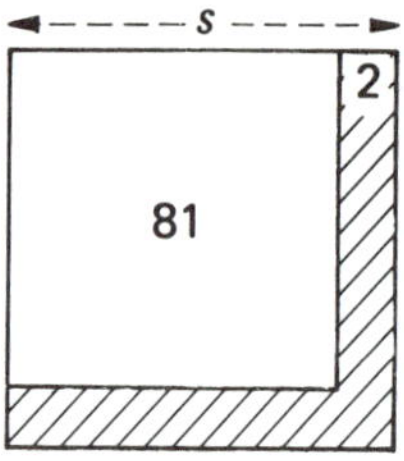


Fig. 51

The solution is immediately obvious, but how is the problem translated into symbols so that this solution is obtained?

Fig. 51 shows that

$$(s-2)^2 = 81$$

from which we have

$$s-2=9$$

$$s=11$$

Now the equation $(s-2)^2 = 81$

can be written as $s^2 - 4s - 77 = 0$

and this form of the equation is similar to that of $t^2 - 20t + 36 = 0$, the equation which originally posed our problem; in each there is the square of the unknown, a first degree term, and a number.

It seems reasonable then to see if our equation

$$t^2 - 20t + 36 = 0$$

can be changed to a form in which there occurs a perfect square expression, analogous to $(s-2)^2$.

This digression has been introduced deliberately at this stage of the discussion to show why 'completing the square' of a quadratic expression in solving an equation is a logical and 'natural' step to take. Teachers will judge for themselves whether with their pupils they should develop the work this way, or whether the steps just illustrated should come prior to the introduction of the topic.

The first two terms of $t^2 - 20t + 36$ can be made into a perfect or complete square by the addition of 100, and to preserve the equality, this number must be added to the right hand side of the equation as well.

Hence

$$t^2 - 20t + 36 = 0$$

$$(t^2 - 20t + 100) + 36 = 100$$

i.e.

$$(t-10)^2 = 64$$

whence

$$t - 10 = 8 \text{ AND } t - 10 = -8$$

so that

$$t = 18 \text{ AND } 2.$$

The object is therefore 176 m above the ground after about 2 s AND after about 18 s.

These values may be checked on the graph, when the readings will again be recognized to be only approximations.

(Note that the solution of the equation quite sensibly gives 2 s and 18 s. Pupils are sometimes puzzled by the conclusion in solving quadratic equations, that, say $x = 2$ *or* 5 as the text-books assert. When such equations are related to real problems, either the two solutions are both correct or one solution will not apply because it is a negative quantity not appropriate to an actual measurement.)

Before proceeding with rather more difficult examples, it will be useful to make a list of the algebraic techniques and manipulations which were involved in the above development, and with which pupils would consequently have to be familiar, before beginning to study quadratic equations in this way.

They are:

(a) knowledge of notation, including indices
(b) substitution in a formula
(c) graph of a formula and its uses
(d) transposition and manipulation of terms in an equation
(e) 'completing the square' (including knowledge of bracket expansions and perfect square expressions)

together with directed number and square root.

This means that quite a formidable programme in algebra has to precede the formal study of quadratic equations and most teachers will be reluctant to impose the full treatment on their less able pupils. Yet the inability to master formal manipulations need not prevent pupils from understanding certain important mathematical principles; stages (a), (b) and (c) above are sufficient to show them that higher degree equations *can* be solved, even if the answers are inexact, and that one aim of mathematics is to develop better and better techniques.

It might be worth noting, as a final reference to the formula

$$h = 98t - 4{\cdot}9t^2$$

that

$$t = 0 \text{ and } t = 20$$

are solutions to the special quadratic equation

$$98t - 4{\cdot}9t^2 = 0$$

as the table calculations have already shown.

After problems such as this one have been used to provide first a motivation and then a method for dealing with the solution of quadratic equations, simple mechanical exercises taken from a textbook now have a purpose even though in themselves they have no concrete application to actual problems. They serve as practice examples through which skill in the method can be developed before the technique is used in more realistic situations.

The use of the graphical introduction in the above discussion will strengthen the conviction of teachers that the graph is a powerful instrument in mathematical investigation. Even if some of our pupils can go no further than to draw graphs of formulae and use them as suggested above, they will begin to realize their usefulness as a tool; they may also be helpful to appreciate in a small way that the very limitations hint at the possibilities of a deeper mathematical treatment than they are as yet capable of making. Meanwhile the graph should be viewed as an important instrument which does not require a wide mathematical background before it can be usefully and interestingly used.

## Conclusion

In concluding this book it must be re-iterated that much of its content is *not* advocated for the majority of pupils who are now under our consideration. Many of the manipulations and processes are needed in further mathematics – especially in calculus – but we are concerned with those for whom the Mathematical Association in its report *Mathematics in Secondary Modern Schools* says 'We ask that the traditional mathematics syllabus be regarded very critically and the purpose for which any of it is included be reviewed'. Here the formula has been taken as the basis for understanding and development, and most of our pupils will gain an appreciation of its use in practical fields through simple illustrations which do not require some of the more formal techniques which have been dealt with. More than this would be a perpetuation of the kind of algebra which has been largely meaningless even to more able pupils, and possibly to some teachers as well. Yet there is material here which will demonstrate the importance of mathematics as a language for describing relationships and then expressing them in varied forms through operations which reveal greater significance and power. From it, the discriminating teacher will be able to select what is appropriate for pupils of mixed ability and attainment, and help them to sense the vital part mathematics plays in science and technology.

## Note on Apparatus

In the book *Building Up Mathematics* (Hutchinson Educational), Dr Z. P. Dienes devotes a chapter to algebraic concepts. The development suggested there is through the use of Algebraic Experience Material (A.E.M.).

This material is obtainable from:

The Educational Supply Association, Ltd,
Pinnacles,
Harlow,
Essex.